SLEEP MEDICINE

MADE RIDICULOUSLY SIMPLE

AF731885

Fabian H. Rossi, MD
Director, Clinical Neurophysiology Laboratory
Orlando VA Medical Center
Associate Professor, Department of Neurology
University of Central Florida Medical School
Orlando, Florida

Contributing Authors
Sherwin Mina, MD
Nina Tsakadze, MD
Elisa Marie Rossi
Elizabeth Gonzalez
Joshua Rossi
Aunali Khaku, MD

Cover by Matteo Farinella
Cartoons by Richard March

MEDMASTER

ISBN# 978-1-935660-41-5

Made in the United States of America
Published by MedMaster, Inc.
P.O. Box 640028

Dedication

To God Almighty, captain of my soul

To His masterpiece, my beloved and beautiful wife Elizabeth

To the pride of my life, my precious children Elisa Marie and Joshua

To my parents in heaven, Aurora and Ruben, and my parents on earth, Ela and Jose Ramon, whose exemplary and moral lives are beacons for me during times of adversity

To my brother Adrian, for his infallible support

Contents

Introduction

Current reference textbooks commonly provide an extensive and in-depth review of sleep medicine and are of limited use to busy students or clinicians. This book is a brief overview. It introduces medical students, physician assistants, nurse practitioners and general practitioners to the basic aspects of sleep medicine in a way that is easy to read and learn, practical, concise, and fun. It emphasizes the pathophysiology, diagnosis, and treatment of the most common sleep disorders encountered in clinical practice.

Sleep disorders come in many forms. Some people have difficulty falling asleep or remaining asleep. Others experience the opposite – daytime drowsiness. The causes may be physical (abnormal functioning of the brain or other body areas), drug-related, or psychological. The treatment of sleep disorders varies, depending on the cause.

Let us begin with a brief overview of the anatomy and physiology of sleep.

1

Anatomy and Physiology of Sleep

Neural Pathways for Arousal

Figure 1-1 Illustrates the ascending reticular activating system, a key pathway associated with arousal. It originates in the gray matter in the brain stem and projects to the brain. There are also accessory pathways that connect the brain stem with a number of regions of the brain. It is not just the actual connections that are responsible for wakefulness and sleep, but also the chemicals associated with the pathways. Some of the chemicals promote wakefulness while others promote sleep.

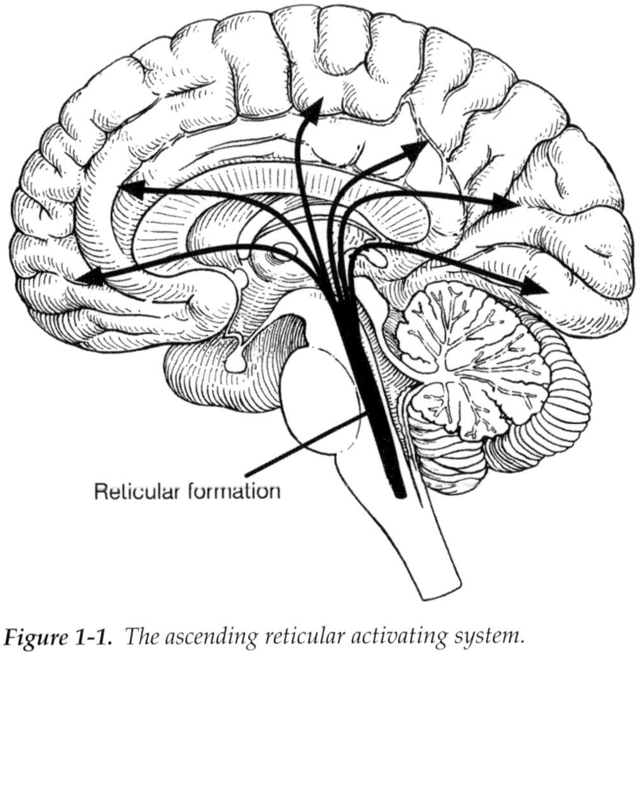

Figure 1-1. The ascending reticular activating system.

For instance, *hypocretin* is an excitatory hypothalamic neuropeptide that promotes wakefulness. Damage to hypocretin-containing neurons causes *narcolepsy*, a sleep disorder characterized by overwhelming drowsiness and sudden episodes of falling asleep during daytime activities. *Melatonin* is a pineal gland hormone that promotes sleep. Melatonin deficiency may result in insomnia.

Damage to the pathways for arousal may cause a state of deep sleep, where it is difficult to awaken the patient.

The Circadian Rhythm and Homeostatic Drives

Wakefulness and sleep are two opposite functions. Wakefulness obviously has the purpose of keeping the person alert, active and functional. Sleep is also important, as can be seen in the consequences of sleep deprivation (**Fig. 1-2**). Sleep functions in a broad sense to "recharge the batteries." Sleep is essential for energy conservation, maintaining body temperature, and sustaining cognitive function (emotions, behavioral health, memory and learning), the immune system, and survival.

Awake and sleep times are controlled by the *circadian rhythm drive*, which primarily promotes wakefulness,

FIGURE 1-2 CONSEQUENCES OF SLEEP DEPRIVATION

- Excessive daytime sleepiness
- Hyperactivity in children
- Mood changes
- Cognitive impairment
- Increased risk for accidents
- Increased risk for infections
- Decreased productivity
- Increased body weight
- Increased sympathetic nervous system activity
- Decreased libido
- Increased mortality

and the *homeostatic drive*, which promotes sleep. The two drives complement each other to maintain a healthy balance of awake and sleep times.

REM and Non-REM (NREM) Sleep

Sleep may be divided into Non-REM (NREM) sleep and REM sleep (REM = Rapid Eye Movements). Non-REM sleep is subdivided into 3 sleep stages: Stage NREM1, Stage NREM2, and Stage NREM3, all seen as different electrical wave activities on an EEG (electroencephalogram).

- *Stage NREM1 sleep*, a bridge between awake and asleep, is a form of light sleep where individuals are easily aroused. It progresses to Stage NREM2.
- *Stage NREM2 sleep* is a deeper type of sleep.
- *Stage NREM3 sleep* ("slow EEG wave sleep") is a deep sleep, where people are hard to arouse.
- *REM sleep* (Stage R), a fourth stage of sleep, is defined by three things:

1. EEG features that resemble those in wakefulness
2. Marked loss of muscle tone (*atonia*)
3. Rapid eye movements

REM may be divided into *phasic REM*, with bursts of eye movements interspersed with *tonic REM*, which lacks rapid eye movements. Tonic REM can still be identified by cortical EEG that mimics the EEG of wakefulness, and decreased or absence of muscle activity as seen on an electromyogram (EMG).

There are "REM-on" and "REM-off" neurons in the brain stem. REM-on neurons promote REM sleep, while REM-off neurons inhibit REM sleep. REM-on neurons have a mutual inhibitory interaction with REM-off neurons.

The neurons that induce non-REM sleep are located in the anterior hypothalamus. They inhibit the ascending reticular activating pathways. Damage to the anterior hypothalamus causes insomnia and sleep fragmentation.

The neurons that generate REM sleep are located in the brain stem's upper pons. Damage to the upper pons neurons causes *REM-sleep behavior disorder*, where loss of muscle tone during REM sleep does not occur, and patients are able to physically "act out" their dreams.

Physiologic Changes During Sleep

During both non-REM and REM sleep, there are physiological changes that normally take place in different organ systems (**Fig. 1-3**).

Dreaming

Dreaming occurs in all stages of sleep, but is more common in REM sleep, where the dreams tend to be more complex and irrational than during NREM sleep, where dreams are more realistic and simpler. The value of dreaming is not firmly established. Dreaming probably represents the conscious perception of images, thoughts, and emotions in some way linked to events while awake. Dreaming affects waking emotions in a positive or negative way depending on the dream content, ability to recall the dream, and dream frequency. Nightmares negatively affect daytime mood. Dreaming also improves daily creativity and problem-solving abilities.

Strokes, traumatic brain injuries, neurodegenerative disorders, brain infection, etc. that affect dreaming tend to decrease the frequency and complexity of dreams.

FIGURE 1-3 PHYSIOLOGICAL CHANGES DURING SLEEP		
Organ System	**Non-REM Sleep**	**REM Sleep**
Cardiovascular System	• Decreased heart rate • Decreased cardiac output • Decreased blood pressure • Decreased coronary and cerebral circulation	**Tonic REM (REM absent)** • Decreased heart rate • Decreased cardiac output • Decreased blood pressure • Increased coronary and cerebral circulation **Phasic REM (REM present)** • Increased heart rate • Increased cardiac output • Increased blood pressure • Increased coronary and cerebral circulation
Respiratory System	• Mild decrease in ventilatory response to hypercapnia and hypoxia • *Periodic breathing* - regular amplitude and frequency • Decreased respiratory rate • Increased arterial carbon dioxide • Decreased arterial oxygen	• Significant decrease in ventilatory response to hypercapnia and hypoxia • *Irregular breathing* –variable amplitude and frequency • Increased respiratory rate • Increased arterial carbon dioxide • Decreased arterial oxygen
Autonomic Nervous System	Compared to wakefulness: • Decreased sympathetic nervous system activity • Increased parasympathetic nervous system activity	Compared to non-REM sleep: • Increased sympathetic activity during phasic REM • Increased parasympathetic activity during tonic REM
Gastrointestinal System	• Decreased swallowing • Decreased esophageal and intestinal motility • Gastric secretion is max between 10pm-2am and lowest between 5-11am	• Same as non-REM
Thermoregulatory System	• Decreased core body temperature	• Same as non-REM
Renal and Genitourinary System	• Increased water reabsorption • Decreased glomerular filtration	• Same as non-REM • Penile tumescence in men and clitoral tumescence in women
Endocrine System	• Increased Growth Hormone • Increased leptin (appetite-decreasing) hormone • Increased prolactin • Increased renin	• Decreased cortisol • Decreased ghrelin (appetite- increasing) hormone • Decreased insulin
Dreaming	• Less vivid	• More vivid, visually intense

2

Evaluation of Sleep Disorders

The diagnosis and treatment of sleep disorders depend on

- an accurate sleep history (from patient, bed partner, roommate, or family member)
- physical examination
- sleep tests

Sleep History

Sleep disorders present with one or more of the following:

A. ***Excessive daytime sleepiness***
B. ***Insomnia***
C. ***Abnormal behaviors during sleep***

A. ***Excessive daytime sleepiness*** means difficulty staying awake and alert during the day. It can be due to *primary* or *secondary sleep disorders.*

Primary sleep disorders that cause excessive daytime sleepiness include

- obstructive sleep apnea (apnea = cessation of breathing) — commonly presents with snoring, breathing interruptions during sleep, and morning headaches
- narcolepsy (a tendency to fall asleep when in relaxed surroundings)
- restless leg syndrome
- circadian rhythm disorders

Secondary sleep disorders that cause excessive daytime sleepiness include

- medical disorders (congestive heart failure, gastro-esophageal reflux, asthma, emphysema, pain conditions)
- neurological disorders, such as dementia, Parkinson's disease, and multiple sclerosis
- psychiatric disorders (e.g. depression and anxiety)
- medications, drugs (including alcohol abuse, coffee intake close to bedtime, and the use or abuse of medications that impair sleep)
- insufficient sleep time (the most common cause. Individuals who sleep consistently less than six hours per night commonly develop excessive daytime sleepiness.)
- using the bed for activities other than sleep or sex

B. ***Insomnia*** can mean difficulty falling asleep (*sleep-onset insomnia*), staying asleep (*sleep-maintenance insomnia*), or *early morning awakening.* Insomnia is often caused by psychiatric disorders, especially depression and anxiety. Patients with depression have difficulty staying asleep and early morning awakening, while patients with anxiety present with sleep-onset insomnia.

The same medical and neurological disorders that cause excessive daytime sleepiness may also cause insomnia. Obstructive sleep apnea and narcolepsy fragment the night sleep and may cause difficulty staying asleep, which results in daily drowsiness. See **Chapter 5** on Insomnia for further information.

C. ***Abnormal behaviors during sleep*** depend on whether they occur during non-REM or REM sleep.

Abnormal behaviors during *non-REM sleep* are common in children and include confusion, sleepwalking, and scare-like reaction. The patients are difficult to arouse during one of these events and have no recollection of it in the morning.

Nocturnal seizures most commonly arise during NREM sleep (**Chapter 8**).

Abnormal behavior *during REM sleep* occurs in older male patients and consists of dream enactments with purposeful movements, such as throwing punches or kicks, running, jumping from the bed, or screaming. The patient often wakes up with full recollection of the dream and has difficulty resuming sleep.

Figure 2-1 lists basic questions during the sleep medicine interview.

Several sleep questionnaires assess different aspects of sleep, such as sleep quality, degree of drowsiness, level of insomnia, risk for sleep breathing disorder, and risk for restless leg syndrome. The most commonly used questionnaire is the *Epworth Sleepiness Scale* **(Fig. 2-2)**, designed to assess daytime sleepiness. It assesses the risk for dozing or falling asleep during eight different sedentary conditions, in contrast to just feeling tired. Each answer uses a *Likert scale* (0,1,2,3) with a maximal score of 24. *A score of 10 or greater suggests Excessive Daytime Sleepiness.*

FIGURE 2-1 KEY QUESTIONS IN EVALUATION OF SLEEP DISORDERS

- What time do you go to bed and get up on weekdays and weekends?
- Do you have problems falling asleep, staying asleep, or awakening in the morning?
- Do you snore, snort, gasp, choke, or stop breathing during sleep?
- Do you have any discomfort in or jerking of your legs at night?
- Do you have any abnormal movements or frequent movements during your sleep?
- Do you doze easily or fall sleep in quiet and monotonous situations?
- Do you take naps?
- What medications do you take?
- Do you consume alcohol, coffee, or illicit drugs?

FIGURE 2-2 EPWORTH SLEEPINESS SCALE

Situation	**Likert Scale (chance of dozing/sleeping)**
Sitting and reading	0 1 2 3
Watching TV	0 1 2 3
Sitting inactive in a public space	0 1 2 3
Being the passenger in a car for an hour with no break	0 1 2 3
Lying down in the afternoon	0 1 2 3
Sitting and talking to someone	0 1 2 3
Sitting quietly after lunch, no alcohol	0 1 2 3
Stopping for a few minutes in traffic while driving	0 1 2 3
Total Score	

Sleep diaries are very helpful in assessing sleep disorders. Patients record their sleep and wake time over 2 weeks, total sleep time during a 24-hour period, arousals at night, and sleep patterns.

Physical Examination

Examination starts with measurement of weight, height, *Body Mass Index* (kg/m^2, where kg is the person's weight in kilograms and m^2 the person's height in meters squared), neck circumference, and blood pressure.

- Overweight individuals (Body Mass Index 25 or greater), especially morbidly obese patients (BMI 40 or greater), are at risk for obstructive sleep apnea.
- A *neck circumference* 16 inches or greater in women and 17 inches or greater in men increases the probability for obstructive sleep apnea.
- *Abnormal facial, nasal, and oral structures* are also risk factors, e.g. *mid-facial hypoplasia* (reduction in nasopharynx space); *retrognathia* (abnormal posterior mandible displacement); *micrognathia* (undersized jaw); *nasal narrowing* caused by congestion, deviated septum, or nasal polyps; and *oral narrowing* by macroglossia (enlarged tongue), posterior tongue displacement, or tonsillar enlargement.

Sensory exam of the lower extremities is warranted when restless leg syndrome is suspected, due to its association with neuropathies (**Chapter 9**).

Sleep Tests

The sleep tests most commonly used to investigate sleep disorders are

- nocturnal polysomnography
- the multiple sleep latency test
- the maintenance of wakefulness test
- actigraphy

The first three tests assess excessive daytime sleepiness, while actigraphy assesses circadian rhythm disorders and insomnia.

Nocturnal Polysomnography

Nocturnal polysomnography is the continuous recording of different physiological parameters that change during awake and sleep, including

- brain activity (on EEG)
- eye movements
- nocturnal movements
- cardiac rhythms
- airflow
- respiratory effort
- body positions
- oxygen saturation in adults
- carbon dioxide levels in children

Nocturnal polysomnography is the main technique used to study sleep disorders, especially sleep-disordered breathing, excessive daytime sleepiness, and, occasionally, unexplained nocturnal awakening and unusual behaviors. It also measures the response to sleep therapy.

Nocturnal polysomnography assesses

- the *sleep architecture* (recording of all sleep stages)
- *sleep continuity* (amount of time asleep and awake during the sleep period)
- the relationship between sleep and arousal in sleep breathing disorders and involuntary leg movements (*periodic leg movement*) during sleep

Scoring in polysomnography is made by studying small segments called *epochs*. Epochs are 30-second intervals on the EEG, each scored as the stage of sleep that occupies more than 50% of the epoch.

A minimum of six hours of recording is recommended. Sleep often starts with a transition from being awake to stage NREM1, briefly followed by stage NREM2, then stage NREM3, then back to stage NREM2, and finally to stage REM, which constitutes a *sleep cycle*.

Figure 2-3 shows the sensors (channels) used during polysomnography. EEG electrode placement is shown separately in **Figure 2-4.**

Electroencephalography, via surface electrodes, records the sum of electrical potentials generated by neurons in the brain cortex. These electrodes are placed on specific skull regions and identified by letters that represent the specific region (F: frontal, C: central, O: occipital); and by numbers to identify the side (*odd* numbers: the left; *even* numbers: the right). The electrodes are arranged in a specific order called *montages* (**Fig. 2-4**). The three essential montages are *frontal, central* and *occipital*. The frontal and central montages are particularly helpful to record sleep, while the occipital montage helps to record

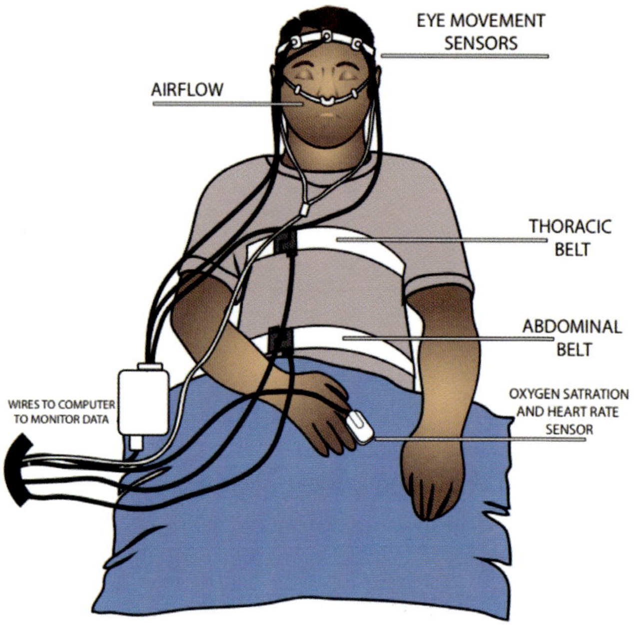

Figure 2-3. *Sensory channels in polysomnography.*

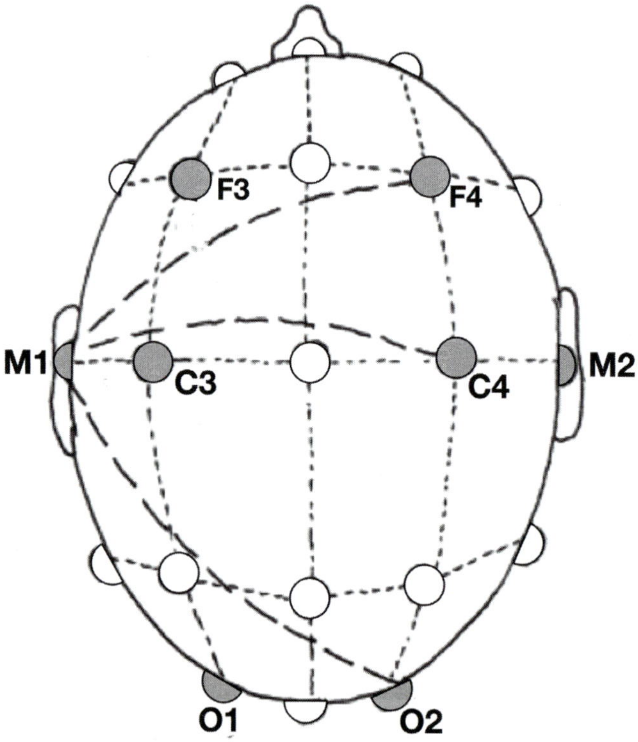

Figure 2-4. *EEG electrode placements.*

awakening. **Figure 2-4** shows the recommended EEG electrode placements.

Figure 2-4. The recommended EEG electrode placements. Right side electrodes (F4, C4 and O2) are paired to create a channel with the contralateral ear electrode (M1, left ear). The difference in electrical potential between these electrodes is amplified by a differential amplifier and then recorded.

Each sleep cycle lasts between 90 to 120 minutes. Most people during an 8hr/night sleep go through 3-5 of these cycles. *Stage NREM3 is more prevalent in the first half of the night, and stage REM during the last half of the night*. The percentage of time spent in each sleep stage compared with the total sleep time is normally 5-10% for stage NREM1, 45-55% for stage NREM2, 10-20% for stage NREM3, and 20-25% for stage REM.

Brain wave frequencies are classified as *alpha, beta, theta,* or *delta* waves (**Figs. 2-5** and **2-6**).

FIGURE 2-5 EEG FREQUENCIES	
Beta	> 13 Hz
Alpha	8-13 Hz
Theta	4-7 Hz
Delta	< 4 Hz

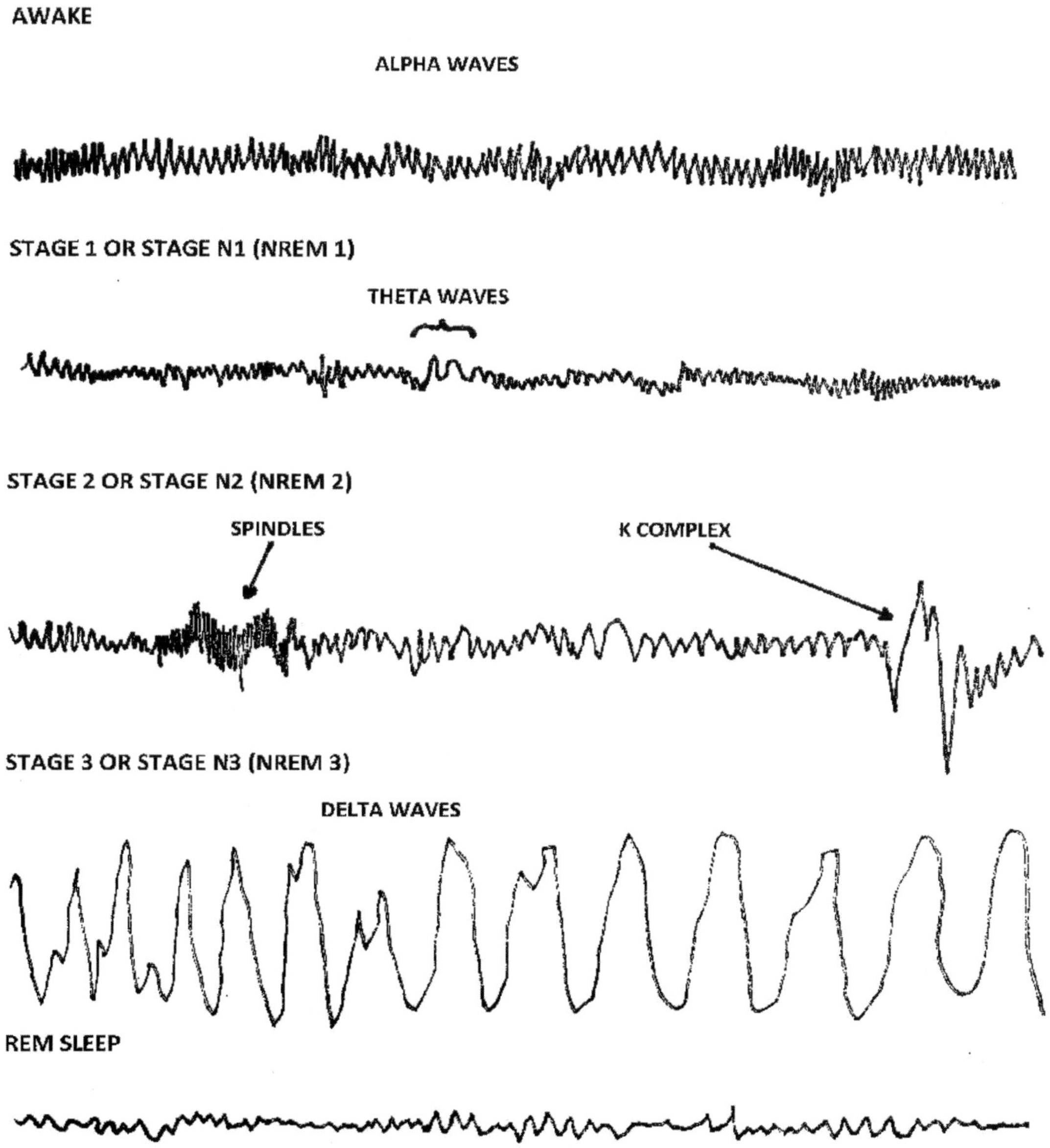

Figure 2-6. *Brain activity frequencies.* Awake state*: alpha activity;* Stage 1 (NREM1; light sleep)*: theta activity;* Stage2 (NREM2)*: spindles (left) and K-complex (right);* Stage 3 (NREM3; deep sleep)*: high delta activity;* REM sleep*: low voltage mixed frequency activity.*

If alpha waves occupy more than half of the epoch duration, the epoch is scored as *stage awake*. However, if alpha waves are reduced to occupy less than half of the epoch, this epoch is scored as *NREM1 sleep (stage N1)*.

Stage 2 sleep, or *stage N2 (NREM2)*, is defined by the presence of two specific waveforms called *spindles* and *K-complexes*. *Spindles* are brief rhythmical sinusoidal and synchronous waves with a frequency of 12 to 15 Hz (1 Hz = 1 cycle/second) lasting > 0.5 seconds. *K-complexes* are sharp slow waves with a duration > 0.5 seconds, predominantly in the central-vertex regions (**Fig. 2-6**).

Stage 3 sleep, or *stage N3 (NREM3)*, is characterized by *high amplitude delta waves* >75 uV in amplitude that occupy more than 20% of the epoch duration (**Fig. 2-6**).

Electrooculography (EOG) records changes in the eye electrical potentials, created by the cornea (positive) and retina (negative) during eye movements. Electrooculography records two types of eye movements: *slow eye movements* during drowsiness and *rapid eye movements* during wakefulness and REM sleep.

Electromyography records muscle electrical activity from the chin and leg muscles to assist in sleep stage determination. Electromyographic activity decreases as sleep deepens and becomes very reduced or lost during REM sleep. *Arousal* from sleep manifests as an increase in electromyography activity. Electromyography also records periodic leg movements and any other abnormal movements during sleep.

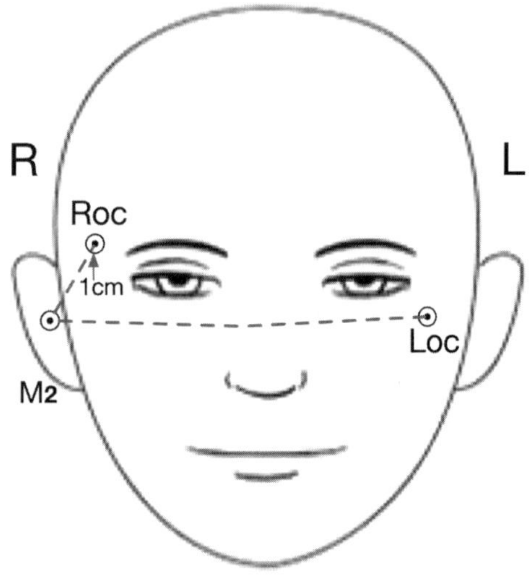

Figure 2-7. *The recommended EOG electrode placements. Electrodes are placed just outside the canthus with the right electrode 1cm above the horizontal midline of the eye and the left electrode 1 cm below the horizontal midline. The two OC (ocular) electrodes connect with the right ear electrode.*

Airflow in polysomography is measured by *thermal sensors* at the nose and mouth, to detect changes in air temperature between warmer expired air and cool inspired air. Also, *pressure sensors* are located at the nose to detect low pressure during inspiration and increased pressure during expiration. Absence of airflow is *apnea*, and reduced airflow is *hypopnea*.

Respiratory effort is measured by recording chest and abdominal movements with inductance plethysmography or esophageal manometry. *Inductance plethysmography* records changes in the chest and abdomen during breathing (**Fig. 2-3**). In *obstructive apnea* there is a paradoxical (opposite direction) movement of the chest and abdomen (contrary to normal). In *central apnea* there are no chest or abdominal movements or effort, and during *mixed apneas* a central apnea is followed by an obstructive apnea (**Fig. 2-8**).

Esophageal manometry consists of placement of a nasogastric pressure sensor in the distal esophagus. This sensor detects respiratory efforts as changes in the pleural pressure that translate into changes in esophageal pressure. The more negative the esophageal pressure becomes, the more intense is the respiratory effort. Esophageal manometry is seldom used because it is an invasive technique and is uncomfortable. It is mainly used in the diagnosis of *upper airway resistance syndrome* (**see Chapter 3**).

Figure 2-8. Airflow and respiratory effort in obstructive, mixed, and central apnea Apnea is a cessation of breathing for more than 10 seconds. In obstructive apnea the respiratory effort continues; in mixed apnea there is a partial respiratory effort during the second half of the apnea; and in central apnea there is no respiratory effort.

Snoring is monitored by miniature microphones. It is used to evaluate obstructive sleep patterns and the relationship between snoring and sleep fragmentation.

Oxygen saturation uses a *pulse oximeter* to estimate the percentage of total hemoglobin that is oxygenated in blood. It is attached to the ear lobule or a fingertip. Normal values are 95-100%. It is primarily used to assess the severity of the sleep breathing disorder, lower oxygen saturation being worse.

The *electrocardiogram* detects cardiac arrhythmias, especially during sleep breathing disorders. The most common arrhythmia observed during obstructive sleep apnea is *bradycardia-tachycardia*.

Body positions are measured by devices or direct observation by the technologist. They are used to study the relationship to sleep breathing disorders. *Obstructive sleep apnea is worse during supine sleep*, especially during REM periods.

Carbon dioxide monitoring is done transcutaneously or through *capnography* (which monitors carbon dioxide concentration in the expiratory air). These techniques are mainly used to assess pediatric sleep breathing disorders and adult hypoventilation (increased carbon dioxide levels of 10 mm Hg or more during sleep versus wakefulness).

Nocturnal polysomnography can be done attended in-laboratory or unattended at home, using portable equipment.

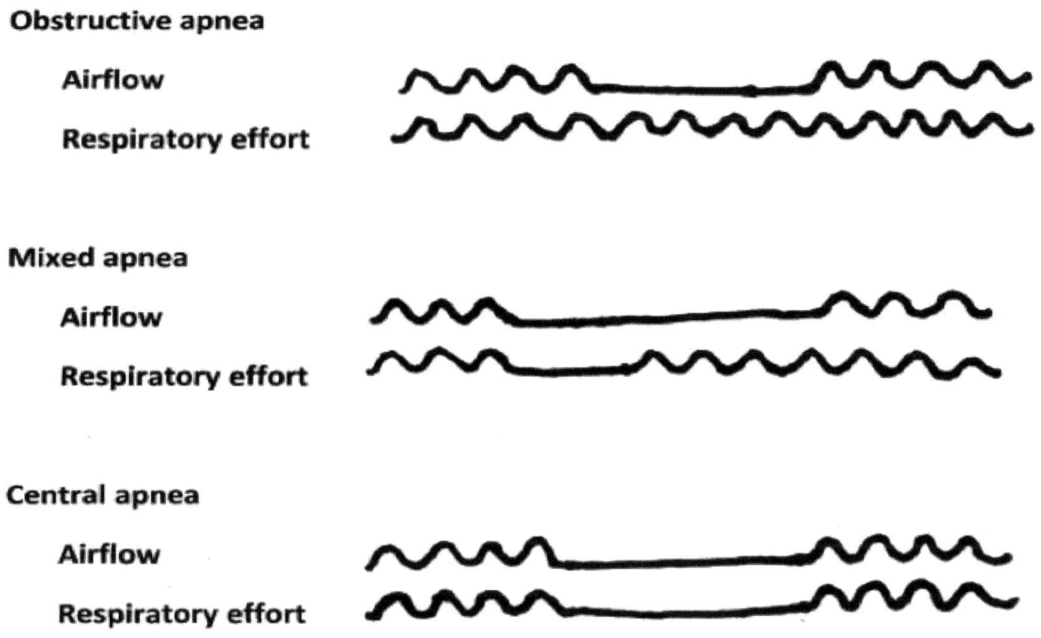

Figure 2-8. *Airflow and respiratory effort in obstructive, mixed, and central apnea.*

FIGURE 2-9 POLYSOMNOGRAPHY IN AWAKE/SLEEP STATES	
Wakefulness (W)	• **EEG:** Alpha activity >50% epoch • **EOC:** Blinking or REM • **EMG:** High muscle activity
Superficial sleep state (NREM1)	• **EEG:** Alpha activity <50% epoch, mixed frequencies, theta activity • **EOC:** Slow, rolling eye movements • **EMG:** Less activity than awake stage
Sleep 2 stage (NREM2)	• **EEG:** Spindles and K-complex • **EOC:** No activity • **EMG:** Less activity than N1
Deep sleep stage (NREM3)	• **EEG:** >20% epoch with delta activity • **EOC:** No activity • **EMG:** Less than N1 and N2
REM sleep (R)	• **EEG:** Low voltage, mixed frequencies • **EOC:** REM during phasic REM • **EMG:** Low or absent
EEG: electroencephalogram; **EOC**: electrooculogram; **EMG**: electromyography	

FIGURE 2-10 NOCTURNAL POLYSOMNOGRAPHY	
Test	**Focuses on:**
Electroencephalography	Brain electrical activity
Electrooculography	Ocular movements
Electromyography	Muscle activity from chin and leg muscles
Electrocardiography	Cardiac electrical activity
Air flow	Movement of air through the nostrils and mouth
Respiratory effort	Chest and abdominal movements
Body position	Supine or lateral positions
Oxygen saturation (used in adults)	Percentage of oxygenated hemoglobin
Snoring	Upper airway vibration
Transcutaneous CO_2 monitoring (used primarily in pediatrics)	Carbon dioxide levels
Capnography (used primarily in pediatrics)	Carbon dioxide levels in outside air
Esophageal manometry	Respiratory effort
Esophageal pH	Esophageal pH levels
Additional EEG channels	Nocturnal seizure activity
Video monitoring	Nocturnal seizure activity and parasomnias (abnormal movements, behaviors, emotions and dreams)

Figure 2-9 summarizes the use of the EEG, EOG, and EMG in the characterization of each sleep state. **Figure 2-10** lists the main tests used to describe sleep architecture and their main uses. **Figure 2-11** defines the main terms used in polysomnography.

Multiple Sleep Latency Test (MSLT)

The *multiple sleep latency test*, mainly used in the diagnosis of narcolepsy, measures the patient's physiological propensity to fall asleep. It is a full-day affair, consisting of four to five 20-minute naps at 2 hr intervals. It measures, with the lights out, how long it takes to fall asleep, whereupon the patient is awakened after 15 minutes. The test is terminated if the patient does not fall asleep within 20 minutes. A nocturnal polysomnography is required the night before the multiple sleep latency test to rule out a sleep breathing disorder and insufficient sleep time (less than 6.5 hours) as the main source of excessive daytime sleepiness.

Two parameters are tested in the multiple sleep latency test: *sleep onset latency* and *sleep onset REM*.

Sleep onset latency is the time elapsed from lights out to the first recorded sleep epoch, normally to stage NREM1. Sleep onset latency of more than 10 minutes is normal; between 8 minutes to less than 10 minutes is inconclusive; and less than 8 minutes is considered abnormal (**Fig. 2-12**).

Sleep onset REM is the recording of REM sleep within 15 minutes from sleep onset. *The combination of short sleep onset latency and two or more sleep onset REMs supports the diagnosis of narcolepsy when clinically suspected.* However, sleep onset latency less than 8 minutes is seen in 30% of the normal population, and sleep onset REM is often seen in very sleep-deprived conditions, drugs, and REM-suppressant drug withdrawal.

Maintenance of Wakefulness Test

The *maintenance of wakefulness test* estimates the person's ability to remain awake and not be overwhelmed by sleepiness while in a quiet and dark environment. It is mainly used where the risk of falling sleep is unacceptable, e.g. piloting a plane. It consists of four 40-minute tests in a relaxing, dimly lit and noiseless bedroom, two hours apart, where the patient is required to remain awake. A sleep latency less than 8 minutes is abnormal, whereas remaining awake in all four trials is the strongest evidence of the individual's ability to stay awake.

Actigraphy

Actigraphy detects movements that represent wakefulness (increased motor activity). Minimal or absent movement represents sleep periods (decreased motor activity). These movements are recorded over days to weeks using an *accelerometer (actigraph)* worn as

FIGURE 2-11 POLYSOMNOGRAPHY DEFINITIONS	
Lights out	Time when recording starts
Lights on	Time when recording ends
Time in bed	Time in bed from lights out to lights on
Total sleep time	Total time from sleep onset to final awakening
Sleep latency	Time in minutes from lights out to first epoch of sleep
REM latency	Time in minutes from sleep onset to first epoch of REM sleep
Awakenings	Number of episodes after sleep onset of > 1 minute of sustained wakefulness – EEG with alpha and beta waves and increase in EMG activity
Wake after sleep onset	Time spent awake after sleep onset
Sleep efficiency	Total sleep time divided by time in bed expressed as a percentage
Arousal	*During non-REM sleep*, an arousal is defined as an abrupt shift of EEG frequencies, including alpha and/or > 16 Hertz (no spindles) for at least 3 seconds. *During REM sleep*, an arousal also requires a concurrent increment in chin EMG-activity.
Movement	Muscle activity that obscures more than half of an epoch so the sleep stage cannot be determined when preceded and followed by epochs with sleep. If muscle activity covers more than half of the epoch preceded and followed by EEG wakefulness, it is not called movement, but rather an awake stage.
Sleep stage percentage	Time in each sleep stage divided by total sleep time
Apnea	Absence or > 90% reduction in airflow by at least 10 seconds and associated with > 3% oxygen desaturation.
Hypopnea	> 30% reduction in airflow by at least 10 seconds and associated with > 3-4% oxygen desaturation
Obstructive sleep apnea	Apnea in the presence of respiratory efforts
Central sleep apnea	Apnea in the absence of respiratory efforts
Mixed sleep apnea	Apnea period that starts with a central sleep apnea and follows with an obstructive sleep apnea
Respiratory effort-related arousals	Arousals associated with reduction in airflow and increased respiratory effort for at least 10 seconds without apnea, hypopnea, or oxygen desaturation
Apnea-hypopnea index	Number of apneas plus hypopneas per hour of sleep
Respiratory disturbance index	Number of apneas plus hypopneas plus respiratory effort-related arousals per hour of sleep
Periodic limb movements	Number of periodic limb movements (see **Chapter 9**) per hour of sleep
Periodic limb movements arousal index	Number of periodic limb movements associated with arousals per hour of sleep

FIGURE 2-12 AVERAGE SLEEP ONSET LATENCY VALUES	
Normal	More than 10 minutes
Inconclusive	From 8 to less than 10 minutes
Abnormal	Less than 8 minutes

a wristwatch. This technique determines sleep patterns, time and duration of sleep, sleep disruption (arousals from sleep), and waking time. Actigraphy is mainly used in the assessment of circadian rhythm disorders and insomnia.

3

Obstructive Sleep Apnea (OSA)

Sleep-related breathing disorders are a group of syndromes characterized by repetitive periods of cessation of breathing during sleep. They are divided into

- obstructive sleep apnea
- central sleep apnea
- hypoventilation syndromes

These syndromes present with chronic sleep deprivation along with excess daytime sleepiness and cognitive problems. Sleep-related breathing disorders can be associated with serious cardiovascular and neurological morbidity and mortality. Sleep studies are required for appropriate diagnosis and treatment.

Obstructive Sleep Apnea

Obstructive sleep apnea (OSA) is characterized by intermittent airflow reduction (*hypopnea*) or cessation (*apnea*) despite an ongoing respiratory effort. The condition is caused by a repetitive partial or complete upper airway collapse and is associated with disruptive sleep arousals, sleep fragmentation, and oxyhemoglobin desaturation.

The *apnea-hypopnea index (AHI)* is the number of episodes of apnea plus hypopnea (each one lasting at least 10 seconds) during an hour of sleep (confirmed by decreased blood oxygenation). The AHI quantifies the severity of the obstructive sleep apnea. A normal AHI is <5; mild is 5-15; moderate is 15-30; severe is >30.

The *respiratory distress index (RDI)* is a summary of apneas plus hypopneas plus *respiratory effort-related arousals (RERAs)* per hour of sleep. RERAs are transient arousals from sleep caused by inspiratory effort required to move air across a narrowed airway without meeting the criteria of hypopnea or apnea. A normal RDI is <5.

Obstructive sleep apnea is the most common of all the sleep-related breathing disorders. It affects 2-4% of the adult population in North America, but is often underdiagnosed. It increases with age and is more common in men than women, probably because body fat distribution associated with neck enlargement is more common in men. After menopause, the incidence becomes closer between the sexes.

Pathophysiology of Obstructive Sleep Apnea

The forces that keep the pharynx open are the intraluminal pressure and the tone of some dilator muscles, such as the *genioglossus* (a tongue muscle), the main pharyngeal dilator muscle. During sleep, neuromuscular tone decreases, including that of the dilator muscles, leading to a decrease in upper airway patency. The forces that make the pharynx collapse are the surrounding pressure generated by surrounding soft tissues (muscle, fat, and connective tissue). If the *transmural pressure* (the difference between intraluminal and surrounding tissue pressure) decreases beyond a *critical point,* the pharynx will collapse, especially when the patient is lying supine. The collapse occurs most notably at the levels of the soft palate (nasopharynx) and tongue (oropharynx). If there is respiratory center instability with a reduction in the motor output of upper airway muscles, this further increases the upper airway obstruction, worsening the obstructive sleep apnea (**Fig. 3-1**).

Risk Factors for Obstructive Sleep Apnea

Obesity, especially when accompanied by enlargement of the neck circumference, is the strongest risk factor for obstructive sleep apnea in males and females. A neck circumference of 17 inches or greater in males and 16 inches or greater in females increases the risk. Severe obesity is not only a risk factor for obstructive sleep apnea, but also for *hypoventilation syndrome*, discussed in Chapter 4.

Craniofacial abnormalities are significant risk factors, especially in children. They consist of a combination of anatomical factors that reduce the cross-sectional area of the upper airway and increase the risk of pharyngeal collapsibility during sleep. They include enlarged tonsils, high and narrow hard palate, elongated soft palate, small chin, forward projection of the upper teeth over the lower teeth (*dental overjet*), tongue enlargement (*macroglossia*), deviated septum, chronic nasal congestion, nasal polyps, enlarged uvula, and abnormal positioning of the maxilla and mandible.

Hypothyroidism and *acromegaly* are both associated with macroglossia and increased tissue mass in the pharyngeal region, which increase the risk for obstructive sleep apnea.

Neuromuscular disorders such as muscular dystrophy, post-polio syndrome, and neuroregenerative disorders such as multiple system atrophy increase the risk for obstructive sleep apnea due to decreased muscle tone.

In children, the main risk factors for obstructive sleep apnea are *adenotonsillar hypertrophy* and *neurological disorders*, such as cerebral palsy with abnormal neural control of the upper airway. Low birth weight carries an increased risk of upper airway collapsibility. **Figure 3-2** summarizes most of the risk factors for obstructive sleep apnea.

The *Mallampati classification* (**Fig. 3-3**) divides upper airway obstruction into four classes, based on the relationship between the tongue and oropharyngeal size. The test is conducted with the patient seated, mouth open and relaxed, and tongue maximally protruded. Mallampati classes III and IV are associated with an increased risk for obstructive sleep apnea.

Clinical Presentation of Obstructive Sleep Apnea

Obstructive sleep apnea presents with diurnal (daytime) and nocturnal (nighttime) symptoms. The *daytime complaints* result from chronic sleep deprivation, the most common complaint being excessive daytime sleepiness, varying from mild drowsiness to severe sleepiness during inappropriate circumstances, such as

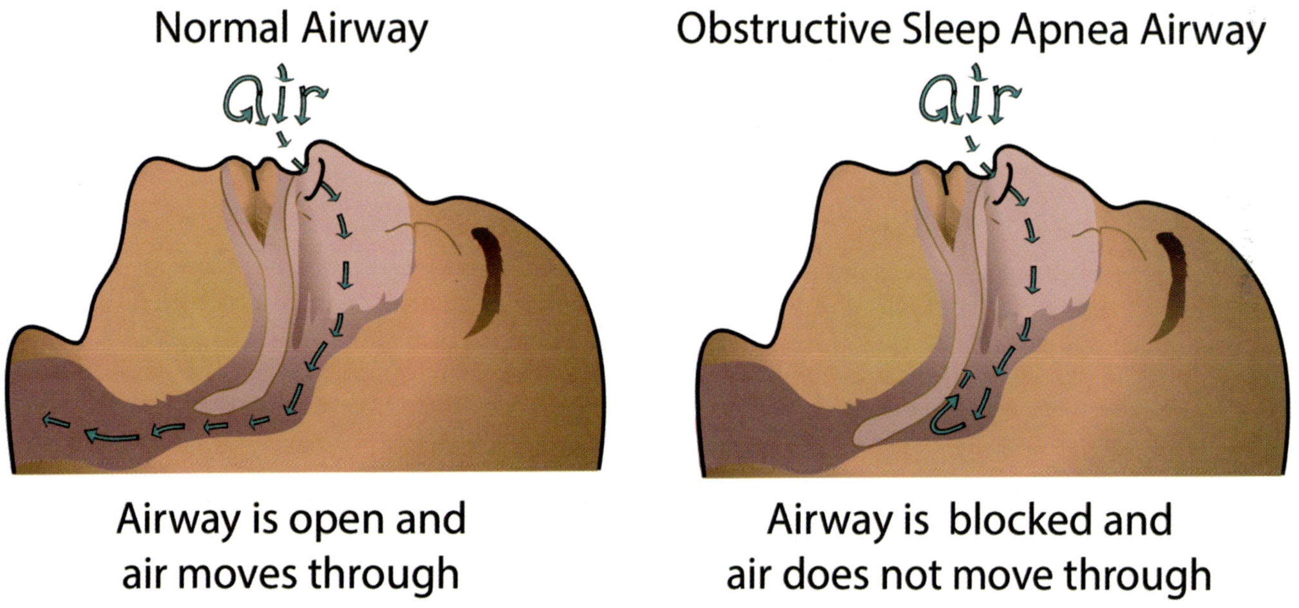

Figure 3-1. Normal airway (left) and airway with obstructive sleep apnea (right).

FIGURE 3-2 RISK FACTORS FOR OBSTRUCTIVE SLEEP APNEA	
Non-Structural Risk Factors	**Structural Risk Factors**
• Obesity (strongest risk factor) • Male sex (2-3 times more common than women) • Advancing age (increases up to the 6th and 7th decade, then plateaus) • Postmenopausal state • Alcohol abuse • Sedative abuse • Opioids • Smoking • Sleeping supine • Family history • Pregnancy • Congestive heart failure • End stage renal disease • Stroke • Chronic lung disease • Atrial fibrillation	• *Retrognathia* (abnormal posterior positioning of the mandible/maxilla) • *Micrognathia* (undersized jaw) • Facial elongation • Posterior facial compression • High arched palate • Mandibular hypoplasia (e.g. *Pierre Robin syndrome*) • Maxillary hypoplasia (e.g. *Down syndrome*) • Hyoid inferior displacement • Adenotonsillar hypertrophy (children) • Nasal obstruction: *congestion, polyps, septal deviation, tumors, trauma, stenosis* • Retro-palatal obstruction: *elongated and posterior placed palate and uvula* • Retroglossal obstruction: *macroglossia and tumor* • Acromegaly • Lower cervical tetraplegia

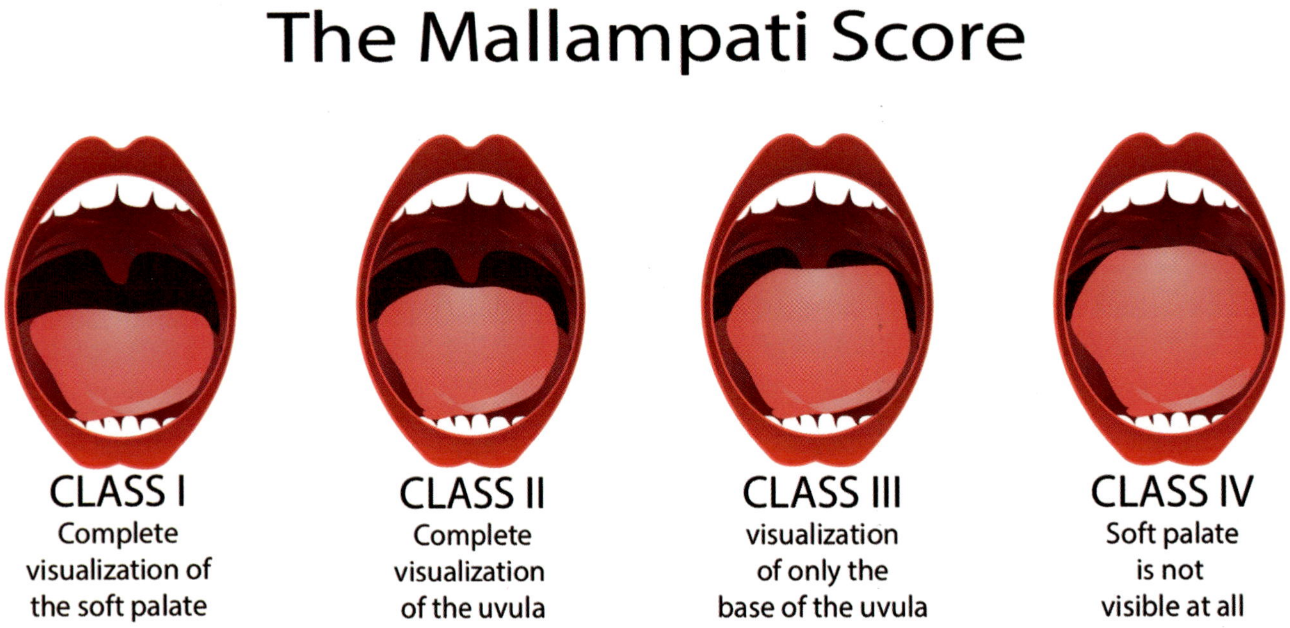

Figure 3-3. Mallampati classification.

driving or having a conversation. Morning symptoms include fatigue, headaches, and lack of a refreshing sleep feeling. Cognitive impairment, inattention, proclivity to accidents, personality changes, depression, decreased libido, and impotence are common.

At night, patients often snore and become agitated and restless. Intermittent apneas may be witnessed by bedmates. Patients start cycles of snoring that progress into apneas, followed by restlessness, gasping for air, and ending in arousals with a snort. Then the whole cycle repeats again throughout the night. **Figure 3-4** summarizes the main clinical presentations of obstructive sleep apnea.

Obstructive sleep apnea worsens with alcohol, sedatives, sleep deprivation, and sleeping supine (greater airway collapsibility). **Figure 3-5** displays the physical findings often seen in obstructive sleep apnea.

FIGURE 3-4 CLINICAL PRESENTATION OF OBSTRUCTIVE SLEEP APNEA

Daytime Symptoms	Nocturnal Features
• Excessive daytime sleepiness • Morning headaches • Fatigue • Inattention • Cognitive decline • Morning confusion • Personality changes • Mood changes • Sexual dysfunction • Propensity to errors	• Snoring • Apneas • Sensation of gasping and choking for air • Insomnia • Nocturia • Enuresis • Sleepwalking

Diagnosis of Obstructive Sleep Apnea

The most used questionnaire in the diagnosis of OSA is the *STOP-BANG questionnaire* (**Fig. 3-6**). It is a very sensitive tool. A lower score (<3) helps to rule out OSA, while higher scores (> or equal to 5) support the diagnosis of OSA. Patients at risk for obstructive sleep apnea (patients with nocturnal snoring and intermittent apnea, daily sleepiness, obesity, and cardiovascular disease) need an *in-laboratory polysomnography*, or in its absence, *home sleep testing* (HST) (portable monitoring of the sleep).

Sleep Studies

Sleep studies are divided based on their complexity into Types I, II, and III (**Fig. 3-7**).

Type I studies are recorded in a sleep laboratory under the supervision of a sleep technologist with full recording of electrophysiological parameters.

Type II consists of an unattended home sleep test (HST) using a minimum of seven channels.

Type III is an unattended HST with a minimum of four channels. Home sleep apnea testing (Types II and III) is intended for sleep apnea screening in patients with high pre-test probability of moderate to severe OSA in the absence of coexistent severe illnesses (severe cardio-respiratory disease, neuromuscular disorders with respiratory involvement, hypoventilation, chronic opioid use, or severe insomnia). HSTs have the pitfall that apneas and hypopneas are unattended; therefore, the study quality may not be as consistent as an in-lab study that is monitored by a sleep technician. If a single and technically adequate home sleep testing is negative or inconclusive, attended in-lab polysomnogram is recommended. Type I testing should be used for those with low pre-test probability (questionable risk) of OSA; coexistence of OSA with severe illnesses; diagnosis of other sleep comorbid conditions or coexisting sleep disorders.

The Apnea-Hypopnea Index (AHI) is primarily used to diagnose and grade the severity of OSA. Most sleep laboratories provide AHI values for NREM and REM and for different body positions. The AHI value is worse during REM when supine (supine favors upper airway collapse, and REM sleep is associated with the lowest muscle tone). In routine polysomnography, respiratory effort-related arousals (RERAs) are not scored (counted); therefore, AHI score alone is reported.

Figure 3-8 shows the diagnostic criteria for obstructive sleep apnea.

Figure 3-9 shows the most important definitions used in the diagnosis of obstructive sleep apnea during sleep studies. The apnea-hypopnea index is used to measure the severity of obstructive sleep apnea.

Figure 3-10 shows how obstructive sleep apnea is divided into mild, moderate, and severe, based on the apnea-hypopnea index. *An AHI lower than 5 is normal.*

Figure 3-11 shows the most important findings encountered in the polysomnography of the patient with obstructive sleep apnea.

The respiratory distress index (RDI) represents the sum of apneas plus hypopneas plus respiratory effort-related arousals (RERAs) per hour of sleep. RERAs consist of increased respiratory effort or decreased inspiratory flow of 10 seconds or longer without meeting the criteria for apnea or hypopnea and followed by an arousal.

FIGURE 3-5 PHYSICAL FINDING IN OBSTRUCTIVE SLEEP APNEA
General Findings
Increased body weight (BMI > 30kg/m^2) Enlarged neck circumference men 17 in or greater (> 43cm) women 16 in or greater (>40cm) Hypertension Depression Congestive heart failure symptoms Pulmonary hypertension Stroke deficits Diabetes-related problems
Upper Airway Findings
Airway size - Mallampati classes III or IV Enlarged tonsils: 3+ or 4+ on a scale of 0 (removed) to 4 Retrognathia Micrognathia Large degree overjet (upper teeth project in front of lower teeth) High-arched hard palate Narrowed lateral airway walls

FIGURE 3-6 STOP-BANG QUESTIONNAIRE HIGH RISK OF OSA: YES 5-8; INTERMEDIATE RISK OF OSA: YES 3-4; LOW RISK OF OSA: YES 0-2	
Do you **S**NORE loudly (louder than talking or loud enough to be heard through closed doors)?	**YES/NO**
Do you often feel **T**IRED, fatigued, or sleepy during the daytime?	**YES/NO**
Has anyone **O**BSERVED you stop breathing during your sleep?	**YES/NO**
Do you have or are you being treated for high blood **P**RESSURE?	**YES/NO**
BMI (Body Mass Index) more than 35 kg/m^2?	**YES/NO**
AGE more than 50 years old?	**YES/NO**
NECK CIRCUMFERENCE *For male*, is your shirt collar 17 inches or greater (>43 cm)? *For female*, is your shirt collar 16 inches or greater (>40 cm)?	**YES/NO**
GENDER male?	**YES/NO**

FIGURE 3-7 SLEEP STUDY SETTINGS FOR OBSTRUCTIVE SLEEP APNEA	
Test Setting	**Test Employed**
Type I **(Done in sleep lab, full recording of all parameters)**	EEG EOG ECG/heart rate Chin EMG Limb EMG Respiratory effort Airflow Pulse oximetry Additional channels for CPAP/BiPAP levels, CO_2, pH, pressure, etc.
Type II **(Home unattended, minimum of 7 channels)**	EEG EOG ECG/heart rate Chin EMG Respiratory effort Airflow Pulse oximetry
Type III **(Home unattended, minimum of 4 channels)**	Respiratory effort Airflow ECG/heart rate Pulse oximetry

FIGURE 3-8 DIAGNOSTIC CRITERIA FOR OBSTRUCTIVE SLEEP APNEA
• Excessive daytime sleepiness not explained by other factors
• Two or more of the following features not explained by other factors: — Choking or gasping for air — Fragmented sleep — Unrefreshing sleep — Daytime fatigue — Impaired cognitive function
• Apnea-hypopnea index (AHI) of five or more episodes per hour

FIGURE 3-9 DEFINITIONS IN OBSTRUCTIVE SLEEP APNEA	
Apnea	Complete cessation of oronasal flow for more than 10 seconds
Hypopnea	Significant reduction (up to 50%) of oronasal flow for more than 10 seconds
Respiratory effort-related arousal	Sequential increase in respiratory effort that leads to an arousal from sleep, but not fulfilling the criteria for apnea or hypopnea
Apnea-hypopnea index (AHI)	Summary of all apneas plus hypopneas per hour of sleep. An AHI < 5 is considered normal.
Respiratory distress index (RDI)	Summary of apneas plus hypopneas plus respiratory effort-related arousals per hour of sleep

FIGURE 3-10 APNEA-HYPOPNEA INDEX CLASSIFICATION OF SLEEP APNEA	
Mild obstructive sleep apnea	$<$ 15 events per hour
Moderate obstructive sleep apnea	15-30 events per hour
Severe obstructive sleep apnea	$>$ 30 events per hour

FIGURE 3-11 COMMON POLYSOMNOGRAPHIC FINDINGS IN OBSTRUCTIVE SLEEP APNEA

- Presence of hypopneas and/or apneas
- Increased arousal frequency
- Increased NREM1 and NREM2
- Decreased NREM3
- Decreased REM
- Paradoxical breathing (chest and abdomen move in opposite directions)
- Oxygen desaturation
- Cardiac arrhythmias

RERA is recorded by the nasal pressure transducers as a flattening airway profile over several breaths followed by the arousal and then restoration of the normal round airflow profile. The gold standard technique to detect RERA is esophageal manometry, which displays a crescendo inspiratory effort for several breaths followed by the arousal. Esophageal manometry is not routinely used in polysomnography because it is uncomfortable. Thermal sensors are not helpful to record RERAs since they only measure absence or presence of airflow.

Peripheral arterial tonometry (PAT) is a more recent technology utilized in the diagnosis of OSA. PAT is a home sleep test, intended for sleep apnea screening in patients with high pre-test probability of OSA. PAT is a portable device that primarily measures peripheral arterial tone and is used with measurements of heart rate, actigraphy, and oximetry to identify respiratory events. PAT is based on the principle that at the end of a respiratory apnea, there is an activation of the sympathetic nervous system that causes vasoconstriction (tonometry) and increased heart rate. This change leads to an attenuation in the PAT signal. PAT follows specific algorithms that diagnose OSA with significant accuracy. PAT can record AHI and RDI and has the advantage of being recorded in-home.

Complications of Obstructive Sleep Apnea

Obstructive sleep apnea activates the sympathetic autonomic nervous system and renin-angiotensin-aldosterone system, which leads to hypertension. In addition, the normal blood pressure, which decreases during sleep by 10% or more of the wakefulness pressure (*dipping effect*) does not take effect or is reduced during obstructive sleep apnea. Non-dipping in combination with hypertension and intermittent hypoxia during obstructive sleep apnea can lead to many cardiovascular complications, including congestive heart failure, coronary disease, cardiac arrhythmias, and stroke. Intermittent hypoxia results in an increase in pulmonary vasoconstriction and pulmonary hypertension, and can lead to right-sided heart failure (*cor pulmonale*). **Figure 3-12** shows the most common complications from obstructive sleep apnea. The *overlap syndrome* is the combination of obstructive sleep apnea with chronic obstructive pulmonary disease (COPD) in the same patient. The patient suffers from even greater hypoxia, nocturnal oxygen desaturation, and pulmonary hypertension than with each individual disorder.

Treatment of Obstructive Sleep Apnea

Treatment of obstructive sleep apnea aims to normalize the apnea-hypopnea index and oxygen saturation and improve sleep efficiency, with resolution of daily symptoms caused by sleep fragmentation (**Fig. 3-13).** Weight loss is recommended in overweight and obese patients. A 10% reduction in weight is associated with a nearly 30% reduction in the respiratory distress. Alcohol should be avoided at least 4 to 6 hours before bedtime; abstinence is preferable. Encourage sleeping on the side rather than supine to reduce respiratory pathway collapse. (Sleeping on the left can also decrease heartburn, while sleeping on the right worsens it.) Exercise programs reduce apnea and hypopnea indexes and improve sleep quality, resulting in a reduction in daily sleepiness.

Optimal positive airway pressure machines decrease the respiratory distress index. This reduces daily drowsiness and depression, and improves cognitive functions, such as memory, alertness, and concentration. It also reduces systemic hypertension, cardiac arrhythmias, congestive heart failure, and the risk of stroke.

Positive Airway Pressure Devices for the Treatment of OSA

Continuous positive airway pressure (CPAP) blows room temperature air into the airway (nose and oropharynx)

FIGURE 3-12 COMPLICATIONS OF OBSTRUCTIVE SLEEP APNEA

- Hypertension
- Pulmonary artery hypertension
- Coronary artery disease
- Cardiac arrhythmias
- Congestive heart failure
- Stroke
- Metabolic syndrome and diabetes mellitus
- Non-alcohol fatty liver disease
- Perioperative complications

to keep the luminal pressure in the pharynx above the pressure of surrounding tissues that tend to collapse the airway (**Fig. 3-14**). CPAP delivers air through a nasal mask, nasal inserts, or full-face mask (for mouth breathers) to keep the airway open (**Fig. 3-15**). It aims to reduce respiratory apneas and hypopneas, and respiratory effort-related arousals. It improves oxygen saturation and sleep efficiency. CPAP delivers a constant pressure (*fixed-pressure continuous positive airway pressure*) regardless of the point in the respiratory cycle (inspiration or expiration).

The two main indications for treatment with CPAP are an apnea-hypopnea index *or* respiratory distress index greater than 15 per hour, with or without symptoms or comorbid disorders. It also is indicated for an apnea-hypopnea index or respiratory distress index between 5 and 14 per hour with any symptoms resulting from obstructive sleep apnea (daily drowsiness, fatigue, insomnia, cognitive impairment, mood changes), or comorbid conditions such as systemic hypertension, cardiovascular disease, chronic obstructive pulmonary disease (COPD), neuromuscular disorders, or stroke.

A full night-attended polysomnography in a sleep lab is the modality of choice to determine the best positive airway pressure requirement. However, *split-night diagnostic*, where the first part of the sleep test is for diagnosis and confirmation of sleep apnea and the second part is for CPAP titration may be also appropriate. *Home auto-titrating equipment* can measure hours of usage and number of respiratory events, estimate the Apnea-Hypopnea Index, and determine the level of pressure required to eliminate most of the abnormal respiratory distress. The pressure commonly used to improve respiration is between 5-15 cm H_2O. However, many patients have problems exhaling against a positive pressure, especially when high pressure is required. This may be relieved with the use of an expiratory pressure relief system; the most commonly used is the *C-flex*, a device that reduces the pressure 1-3 cm H_2O during exhalation, returning to the original pressure during inspiration.

Bilevel positive airway pressure (BiPAP) delivers two types of independent positive pressures: inspiratory positive airway pressure and expiratory positive

FIGURE 3-13 TREATMENT OF OBSTRUCTIVE SLEEP APNEA

General	Weight loss Avoid alcohol Avoid tranquilizers or sleeping pills Avoid supine position
Medications	Topical nasal corticosteroids *Modafinil* (reduces daytime sleepiness) *Armodafinil* (reduces daytime sleepiness) *Dronabinol* (cannabinoid) lowers AHI in moderate to severe OSA
Positive airway pressure machines	CPAP (Continuous Positive Airway Pressure) BiPAP (Bilevel Positive Airway Pressure) APAP (Auto-Positive Airway Pressure) ASV (Adaptive Servo-Ventilation)
Oral appliances	Mandibular repositioning appliances Tongue-retaining devices
Surgery	Nasal procedures Uvulopalatopharyngoplasty (UPPP) Craniofacial reconstruction Tracheostomy

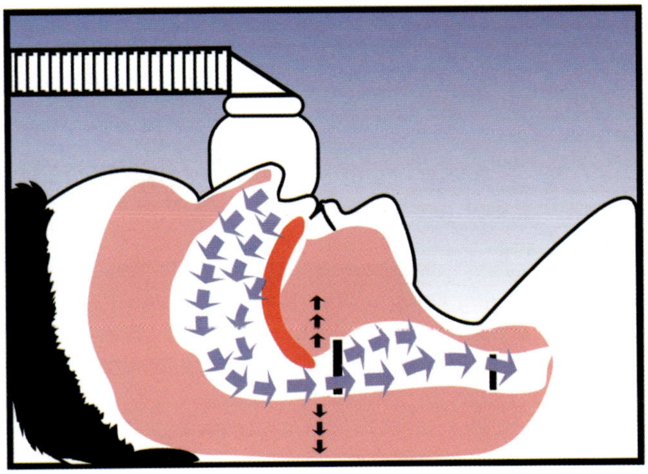

Figure 3-14. CPAP keeps the airway open.

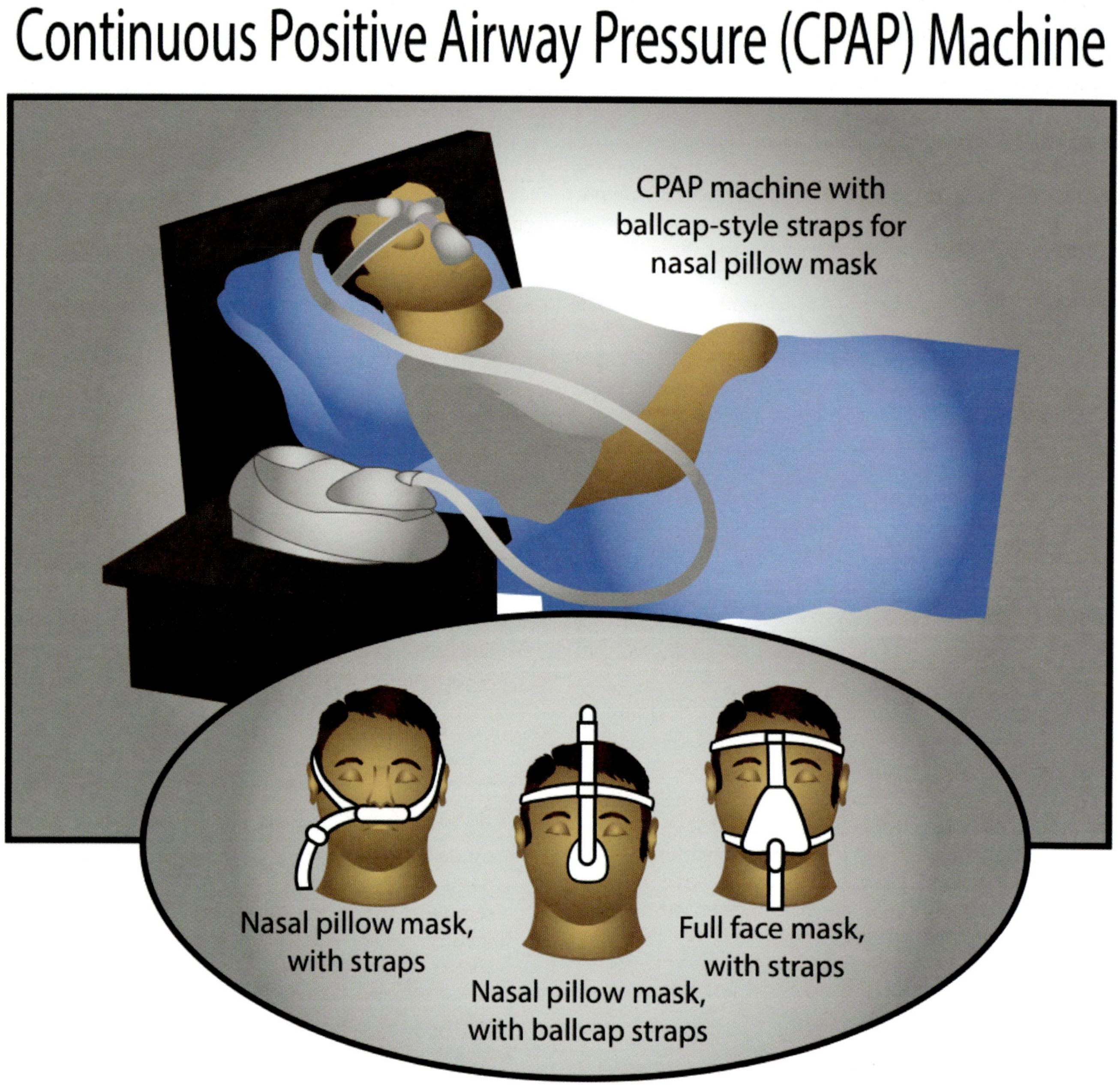

Figure 3-15. *Masks used for continuous positive airway pressure.*

airway pressure, rather than fixed pressure. This form of therapy allows pressure adjustments during inspiration and expiration, with less pressure requirement during expiration, compared with fixed-pressure continuous airway pressure. Bilevel positive airway pressure is recommended in patients who require high fixed pressure but have difficulty exhaling against high pressure. BiPAP is also used to treat the *overlap syndrome* (obstructive sleep apnea + chronic obstructive pulmonary disease) and when OSA is associated with congestive heart failure, neuromuscular disorders, hypoventilation, or central sleep apnea.

Auto-positive airway pressure (APAP) (also called *automatic* or *auto-titrating positive airway pressure*) delivers a fluctuating, but adequate, pressure to maintain an open airway. It responds accordingly to feedback information from airway flow, pressure, and snoring sensors when they indicate airway obstruction. APAP follows algorithms that increase the air pressure whenever an adverse respiratory event is detected (snoring, hypopnea, or apnea). After a therapeutic pressure is obtained, pressure decreases until another airway obstruction is detected. It works similarly to a thermostat that keeps the temperature within a specific range, but focuses on providing an adequate pressure range to maintain an open airway. APAP is mainly used to treat uncomplicated moderate to severe obstructive sleep apnea.

Auto-bilevel positive airway pressure (Auto-BiPAP) uses specific algorithms to provide adequate inspiratory and expiratory positive airway pressures, based on feedback information about upper airway obstruction. Despite auto-titrating, APAP and Auto-BiPAP have not improved patient compliance with the device, as compared to continuous fixed pressure.

Adaptive servo-ventilation (ASV), discussed in Chapter 4, is mainly used when central sleep apnea coexists with obstructive sleep apnea. The device detects periods of decreased airflow during breathing and adjusts immediately by increasing air pressure to support the breathing at the time of decreased flow.

Unfortunately, despite the clear benefit of positive airway pressure therapy, one-third of patients do not use the machine. Of the two-thirds of patients who do use it, a significant number do not use it regularly or the whole night. Common reasons for CPAP failure are non-compliance, inappropriate pressure settings (too much or too little pressure), increased body weight (increased pressure requirements), use of medications that aggravate sleep apnea, and coexistence of other sleep disorders (restless leg syndrome, narcolepsy, etc.). Studies show that if patients are not compliant in the first two weeks, they will not likely be compliant in the long run. Continuous surveillance, from medical staff, and behavioral therapy may help to improve compliance.

Common complications from continuous positive-pressure airflow are

- conjunctival irritation from air leaks around the mask (improve with appropriate fitting of the mask)
- nasal irritation with congestion, rhinorrhea, bleeding from using dry air (improve air humidification; use nasal spray with saline for dryness; use corticosteroids for congestion and ear-nose-throat consultation if required)
- claustrophobia caused by the nasal and facial mask applied tied on the face. This improves with behavioral therapy.
- discomfort and inability to exhale against the air pressure (one can use a system that allows temporary reduction in pressure until the patient falls sleep and then increases the pressure).
- skin abrasions caused by the mask (may require a change of mask type).
- air leak through the mouth (may require a chin strap, a band under the chin to keep the mouth closed. A nasal-oral mask can be also considered).

Oral Appliances for OSA

Oral appliances are indicated in patients with mild to moderate obstructive sleep apnea who prefer oral appliances rather than positive airway pressure devices. Oral appliances are not used in severe obstructive sleep apnea unless other therapies have already failed. Oral appliances aim to enlarge the upper airway and decrease its collapsibility. There are two kinds of oral appliances:

- mandibular repositioning appliances (move the mandible forward)
- tongue-retaining devices (move the tongue forward)

Oral appliances have to be fitted by a specially trained dentist. Complications are temporomandibular joint problems and dental misalignment.

Surgery for OSA

Surgical treatment is indicated for patients with obstructive sleep apnea who fail positive airway pressure therapy and oral appliances. Surgery procedures to correct anatomical obstructions include

- uvulopalatopharyngoplasty (the most commonly performed surgery in adults). It consists in removing uvula, tonsils, and part of the soft palate.
- tonsillectomy with/without adenoidectomy (the treatment of choice in children)
- maxillo-mandibular osteotomy and advancement (**Fig. 3-16**)
- tracheostomy (for severe obstructive sleep apnea associated with life-threatening cardiac complications)

Other effective surgeries are *abdominal surgery* to lose weight and the implantation of a *hypoglossal nerve stimulator* to stimulate the dilatation of pharyngeal muscles. The last device is called Inspire Therapy and consists of a generator [neurostimulator] implanted under the skin of the chest, a sensing lead that detects breathing pattern abnormalities, and a stimulation lead to stimulate key air muscles to keep the airway open during sleeping. Inspire reduces apneas, oxygen desaturation, and daytime drowsiness. It is approved to treat moderate to severe obstructive sleep apnea in individuals 18 years and older.

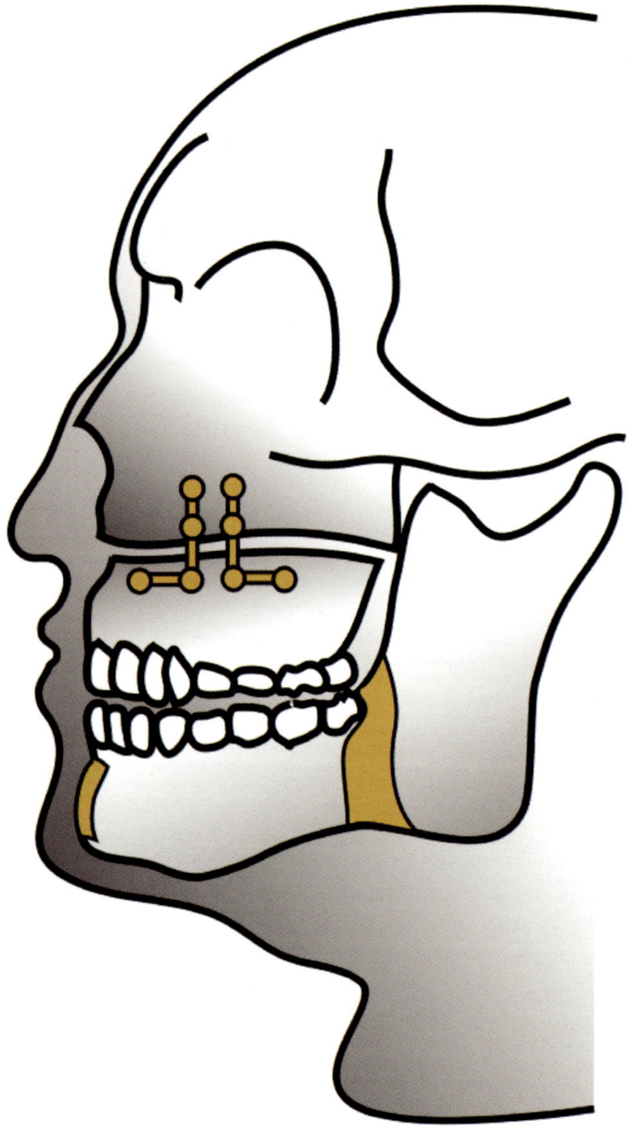

Figure 3-16. *Maxillo-mandibular osteotomy and advancement.*

Other Therapies for Snoring and OSA

Slumber-bump is used in positional OSA. It consists of a sleep belt wrapped around the chest with a hard object on the back. It prevents patients from sleeping on the back and forces them to sleep on the side. It is an over-the-counter device that relieves snoring and obstructive sleep apnea, conditions that worsen when supine.

Humidifiers moisturize the air and reduce the airway congestion caused by dry air. The addition of peppermint, eucalyptus, or lavender oils to the humidifier might have an anti-inflammatory effect that helps with breathing.

The *Winx Sleep Therapy System* (experimental) consists of a mouthpiece with a flexible bar located over the back of the tongue that generates a negative pressure that moves the soft structures forward and enlarges the upper airway.

Provent Sleep Apnea Therapy is an experimental device that aims to increase the expiratory pressure and maintain an open airway. It does this by inserting a nasal cannula in each nostril that allows the air to enter freely while resisting the exit of air.

Upper Airway Resistance Syndrome

Upper airway resistance syndrome is a sleep breathing disorder caused by airway narrowing, elevated airway resistance, and reduced airflow in inspiration. It presents with respiratory effort-related arousals (RERAs) that disrupt sleep and cause chronic fatigue and insomnia, usually less severe than in obstructive sleep apnea. Polysomnography may show sleep fragmentation and an apnea-hypopnea index less than 5 per hour. Intra-esophageal balloon recording is the standard method of diagnosis; it detects a rise in end-inspiratory negative pressure. Treatment is similar to that in obstructive sleep apnea.

Snoring

Snoring results from vibration of the soft palate, uvula, and pharyngeal walls created by the turbulence of the inspiratory air during sleep with a narrow upper airway. Snoring is more common in men and its incidence increases with age. Supine position, obesity, and nasal obstruction increase the risk for snoring. Snoring may occur in isolation or in association with upper airway resistance syndrome or obstructive sleep apnea. Treatment focuses on avoiding alcohol and smoking, which aggravate snoring, losing body weight, and using nasal decongestant sprays. Some people benefit from nasal strips that open the nasal passages or a simple chin strap to keep the mouth closed during sleep. Positive airway pressure therapy is recommended when snoring coexists with upper airway resistance syndrome or OSA. Uvulopalatopharyngoplasty may reduce or eliminate snoring. Radiofrequency tissue ablation has been used to shrink soft tissue to reduce obstruction by the tongue, soft palate or nose. Hypoglossal nerve stimulation may help to bring the tongue forward to reduce airway blockage.

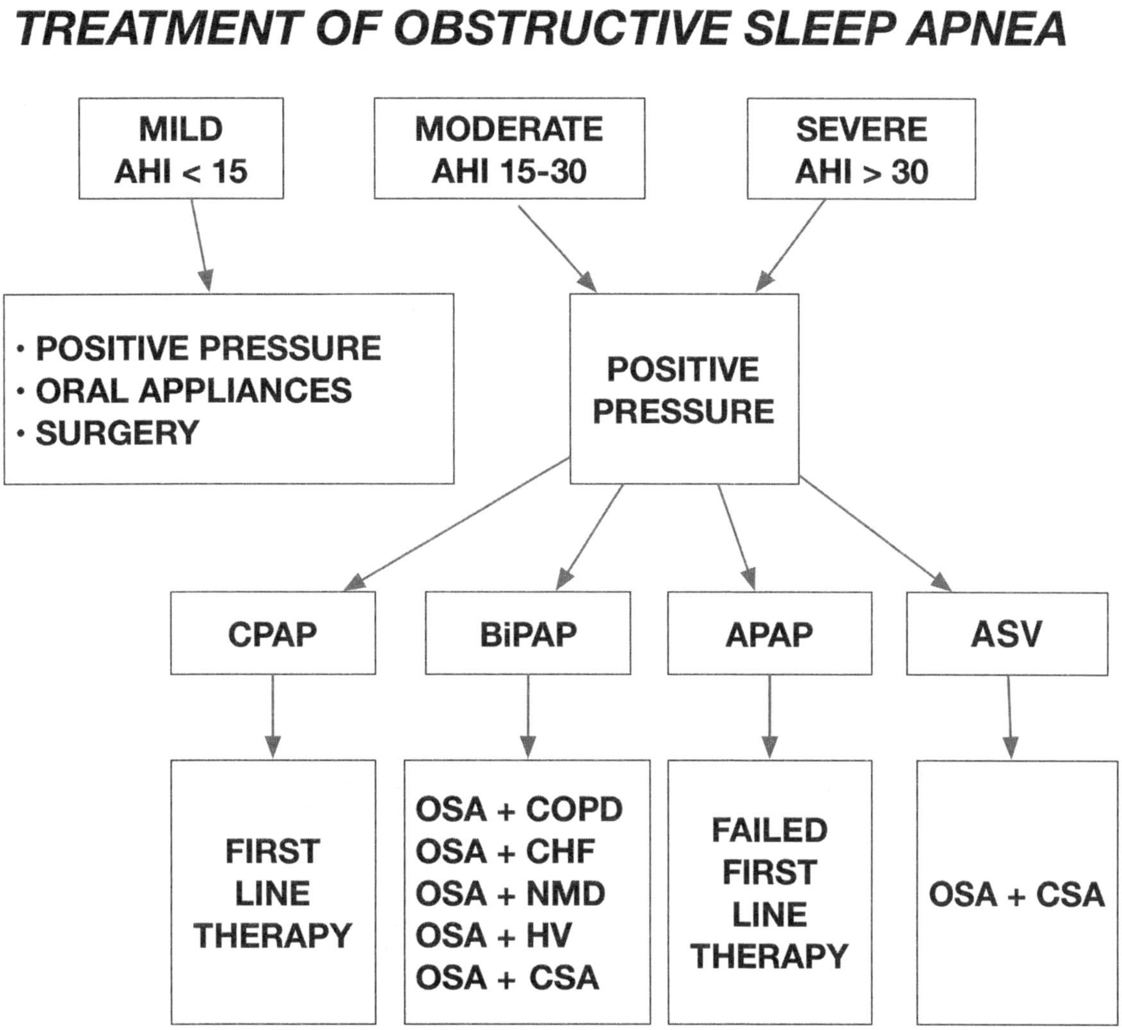

AHI = apnea-hypopnea index
APAP = auto-positive airway pressure
ASV = adaptive servo-ventilation
BiPAP = bilevel positive airway pressure
CHF = congestive heart failure
COPD = chronic obstructive pulmonary disease
CPAP = continuous positive airway pressure
CSA = central sleep apnea
HV = hypoventilation
NMD= neuromuscular disorders
OSA = obstructive sleep apnea

Figure 3-17. *Summary of the treatment approach to obstructive sleep apnea.*

4

Central Sleep Apnea and Hypoventilation Syndromes

Central Sleep Apnea

Central sleep apnea syndrome (CSA) is a repetitive cessation of airflow for at least 10 seconds or more in the absence of respiratory effort, due to an intermittent failure of the nervous system to drive breathing. The symptoms resemble those of OSA, but the mechanisms differ. While OSA is a mechanical problem, CSA is a nervous system problem. CSA, which represents about 10% of the sleep breathing disorders, primarily affects men older than 60 years.

Central sleep apnea may be *idiopathic* or *secondary* to another illness (the most common type). There are some central sleep apneas with high wakefulness body carbon dioxide levels (*hypercapnic*) and those that are *non-hypercapnic* (the most common type).

Hypercapnic-related central sleep apneas occur in the setting of alveolar hypoventilation with daily hypercapnia that worsens during sleep. They result from impaired respiratory output due to respiratory center damage or respiratory organ impairment, such as nerve and/or muscle disorders. These apneas include

- impaired central drive
- impaired respiratory motor control

Impaired central drive ("Won't Breathe") is caused by lesions to the brain stem (e.g. trauma, tumors, specific genetic disorders) that decrease the respiratory response to hypercapnia. *Chronic narcotics usage* affects the central sensors for breathing, leading to central sleep apnea, hypoxemia, and hypoventilation. *Congenital central hypoventilation syndrome* and *obesity hypoventilation syndrome* are discussed below in *Hypoventilation*.

Impaired respiratory motor control ("Can't Breathe") includes several disorders that affect chest wall mobility (trauma to the spinal cord, degenerated anterior horn cells of the spinal cord in *amyotrophic lateral sclerosis* or *post-polio syndrome*), nerve problems (diabetes), muscle issues (*muscular dystrophy, acid maltase deficiency*), and malfunctioning of the neuromuscular junction (*myasthenia gravis*). Breathing deficits may also result from chest wall abnormalities that impair chest contractibility, such as severe *kyphoscoliosis* and trauma to the chest cavity.

Non-hypercapnic related central sleep apneas occur in the setting of normal or low blood carbon dioxide levels. There are several forms of non-hypercapnic central sleep apnea:

- Cheyne-Stokes breathing
- idiopathic central sleep apnea
- high altitude central sleep apnea

Cheyne-Stokes breathing is a crescendo and decrescendo respiratory pattern separated by periods of apnea or hypopnea (**Fig. 4-1**). It may occur during wakefulness or during sleep. It is associated with increased mortality. This respiratory pattern results from an unstable respiratory center with increased chemo-sensitivity to

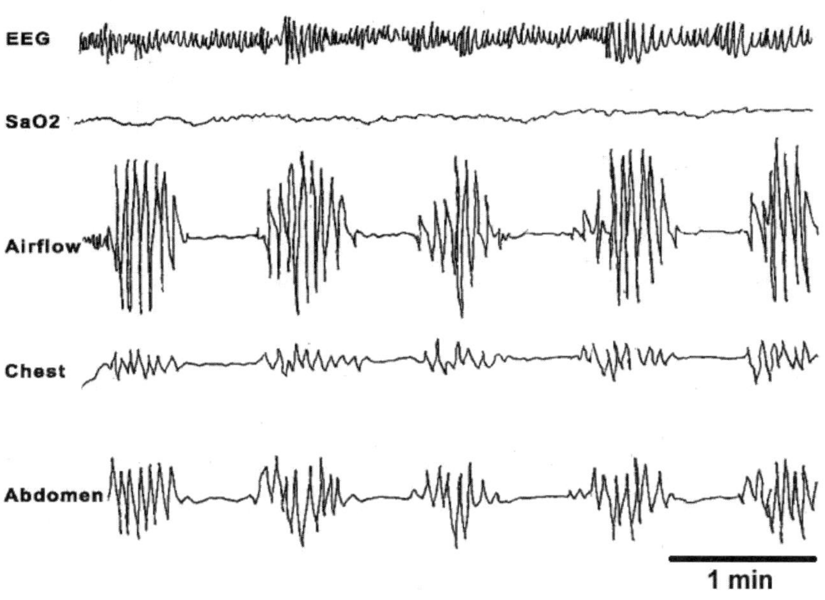

Figure 4-1. Cheyne-Stokes breathing with the crescendo-decrescendo airflow pattern.

hypercapnia, and unstable respiratory control during the transition from wakefulness into sleep.

During the transition from awake to sleep, the carbon dioxide levels fall below the sleep apnea threshold, resulting in an apnea. The subsequent accumulation of carbon dioxide levels then reaches the apneic threshold, and the respiratory effort resumes with progressively increased breathing to the point of overshooting, with a drop in the carbon dioxide, leading again to apnea. The cycle repeats itself. This breathing pattern occurs during light sleep (stages NREM1 and NREM2) and disappears during REM sleep. Arousals occur in the middle of the cycle during maximal respiratory effort. The cycle time lasts 60-90 seconds.

Cheyne-Stokes breathing is most commonly seen in congestive heart failure with ventricular systolic dysfunction (ejection fraction < 45). It is also seen in neurological disorders and renal failure.

Physiologic changes during cardiac failure can trigger this form of respiration. Patients with Cheyne-Stokes respiration present with fragmented sleep, fatigue, daily drowsiness, and symptoms of heart failure, including *paroxysmal nocturnal dyspnea* and *orthopnea*.

Idiopathic central sleep apnea is a rare sleep disorder of unknown cause, more common in males, during light sleep. It may resemble Cheyne-Stokes breathing, with a crescendo and decrescendo respiratory pattern, but with shorter cycles of 20-40 seconds and arousals at the end of the apnea as opposed to the middle of the respiratory effort, as occurs in Cheyne-Stokes. It results from an increased respiratory response to elevated carbon dioxide. Patients present with insomnia or excessive daytime sleepiness.

High altitude central sleep apnea manifests as *periodic breathing,* where clusters of breaths are separated by periods of apnea. It is caused by ascension to high altitudes (8000 feet, or 2500m).

Diagnostic Criteria for Central Sleep Apnea

The patient should display at least one of the following findings to make the diagnosis of central sleep apnea on polysomnography (**Fig. 4-2**).

Treatment of Central Sleep Apnea

Figure 4-3 shows the most common treatments for central sleep apnea.

The treatment approach in central sleep apnea involves treating the underlying disorder and eliminating all potential triggers. Asymptomatic central sleep apnea usually does not require treatment, but does need surveillance. If central sleep apnea becomes symptomatic and does not respond to the above measures, positive airway pressure is warranted.

Continuous positive airway pressure (CPAP) (discussed in Chapter 3) is used in Cheyne-Stokes breathing

FIGURE 4-2 POLYSOMNOGRAPHIC CRITERIA FOR CENTRAL SLEEP APNEA

- 5 or more central sleep apneas and/or central hypopneas per hour of sleep
- The number of central sleep apneas and/or central sleep hypopneas need to be more than 50% of the total number of apneas and hypopneas.

FIGURE 4-3 TREATMENT OF CENTRAL SLEEP APNEA

- Treat the underlying disorder.
- Discontinue narcotics.
- Medications: Acetazolamide and theophylline are respiratory stimulants that reduce the central apnea.
- Oxygen therapy stabilizes the breathing pattern in Cheyne-Stokes breathing.
- Positive Airway Pressure Therapy
  - — CPAP
  - — BiPAP
  - — Adaptive servo-ventilation
- Phrenic nerve stimulator

to reduce central sleep apnea and hypopnea. It also improves cardiovascular function.

Bilevel positive airway pressure (BiPAP) (discussed in Chapter 3) is primarily used in hypoventilation-related central sleep apnea such as the *obesity-hypoventilation syndrome* and *brain stem lesion-related central sleep apneas.* It may contain a backup rate to provide additional breath where a prolonged central sleep apnea is detected. It may also be used in Cheyne-Stokes and idiopathic central sleep apnea.

Adaptive servo-ventilation is set to deliver two types of positive pressure: expiratory positive airway pressure, a fixed-pressure aiming to eliminate obstructive sleep apneas; and inspiratory positive airway pressure that is activated upon detection of central sleep apnea. Inspiratory positive airway pressure is programmed to be equal or greater than the positive expiratory pressure. Adaptive servo-ventilation continuously adapts to the patient's breathing pattern and is thereby more comfortable than CPAP or BiPAP. Adaptive servo-ventilation is mainly used for Cheyne-Stokes breathing, where it reduces sleep apnea and hypopnea and improves arterial oxygen saturation and daily symptoms. It is also used in central sleep apnea-related narcotics usage and complex sleep apnea.

Complex sleep apnea is a sleep breathing disorder characterized by repeated non-hypercapnic central sleep apneas that emerge after eradication of obstructive sleep apneas with positive airway pressure. This probably occurs because of unmasking coexistent central apneas that become more evident as obstructive sleep apneas are controlled by the CPAP. Adaptive servo-ventilation is not recommended in patients with moderate to severe central sleep apnea with symptomatic heart failure and ejection fraction of 45% or less, due to increased mortality.

Increased dead space. Using devices such as a bag or a cylinder attached to a facial mask may increase the content of carbon dioxide in the breathing air and then maintain the carbon dioxide levels above the apneic threshold.

Oxygen supplementation is used to treat high altitude central sleep apneas, where it reduces apnea/hypopnea and improves sleep architecture. Oxygen is also used in Cheyne-Stokes breathing to reduce the hypoxic drive toward hyperventilation. Servo-ventilation is more effective than oxygen supplements to treat Cheyne-Stokes breathing.

Medications seem to have a modest role in non-hypercapnic related central sleep apneas. *Acetazolamide* is a carbonic anhydrase-inhibitor that causes metabolic acidosis and acts as a respiratory stimulant. It has been used in Cheyne-Stokes breathing and in idiopathic and high altitude central sleep apnea. The sedative *zolpidem* (**see Chapter 5**) may reduce the respiratory center's instability.

A *phrenic nerve stimulator* is reserved for patients with central sleep apnea who fail all other therapies. It consists of an implantable device (*neurostimulator*) that sends impulses to the phrenic nerve to induce diaphragmatic contractions, mimicking normal breathing.

Hypoventilation Syndromes

Alveolar hypoventilation results from insufficient ventilation of the pulmonary alveoli, which leads to an elevated arterial carbon dioxide, above 45mm Hg (*hypercapnia*). It is more common in older men, although primary hypoventilation may start during infancy.

The most common cause of alveolar hypoventilation is *chronic obstructive pulmonary disease.*

Hypoventilation worsens during sleep because the minute ventilation declines further during non-REM sleep and reaches its lowest during REM sleep.

Figure 4-4 lists the types of hypoventilation syndromes.

Nocturnal hypoventilation symptoms consist of poor sleep quality, morning headaches, excessive daytime sleepiness, cognitive dysfunction, fatigue, nightmares, and enuresis.

Idiopathic hypoventilation syndrome is caused by reduced chemo-responsiveness to carbon dioxide. There is no abnormality in the respiratory system, neuromuscular system, or chest wall movements, or

FIGURE 4-4 TYPES OF HYPOVENTILATION SYNDROMES

- Idiopathic hypoventilation syndrome
- Congenital central alveolar hypoventilation syndrome
- Hypoventilation due to pulmonary or vascular pathology
- Hypoventilation due to neuromuscular and/or chest wall abnormalities
- Obesity hypoventilation syndrome
- Opioid medications

medications. Its onset occurs in early adulthood, but may start during infancy.

Congenital central alveolar hypoventilation, also known as *"Ondine's curse,"* is a rare inherited disorder of the newborn associated with severe alveolar hypoventilation during sleep. This syndrome is caused by failure of the automatic breathing control (central and peripheral chemoreceptors) to respond to oxygen and carbon dioxide levels. It presents with shallow breathing, apnea, cyanosis, pulmonary hypertension, cor pulmonale, failure to grow, mental retardation, and premature death. Most cases are caused by a mutation of the *PHOX2B gene*. It is associated with *Hirschsprung's disease* and neural crest tumors. The heterozygous form may reach adulthood. Patients require ventilator support for life.

Hypoventilation due to pulmonary and vascular pathology syndrome is caused by diseases that affect the lung parenchyma (interstitial lung disease and pulmonary hypertension), obstruct or increase the resistance in distal airways (emphysema, asthma, and chronic bronchitis), or impair lung vasculature, such as sickle cell disease. These disorders impair gas exchange at the alveoli, resulting in hypoventilation.

Hypoventilation due to neuromuscular and/or chest wall abnormalities decreases the "ventilator pump capacity" through impaired contractility due to weakness of the diaphragm, intercostal, and accessory muscles, as seen in myopathies and amyotrophic lateral sclerosis. Or chest wall elasticity may be limited, as in severe chest deformities (e.g. *scoliosis*).

Obesity hypoventilation syndrome is characterized by the combination of obesity with hypoventilation-induced hypercapnia and hypoxemia. It is worse during sleep. Most patients also have obstructive sleep apnea. They present with symptoms of obstructive sleep apnea and hypoventilation. Hypoventilation becomes complicated with pulmonary hypertension, cor pulmonale, polycythemia, and cardiac arrhythmias.

Hypoventilation caused by opioid medications. Opioids impair the control of central respiratory centers.

Treatment of Hypoventilation

First-line therapy consists of treating the underlying condition, the use of medication to open the airway (bronchodilators), respiratory stimulants to improve ventilation (limited effect), weight loss (surgery may be considered in patients with body mass index > 35 kg/m^2 and obesity hypoventilation syndrome). Oxygen therapy is used to keep arterial oxygen saturation between 90-94%. Too much oxygen is not warranted as it may worsen hypercapnia due to oxygen's effect on suppressing ventilation. **Figure 4-5** summarizes the different treatments for hypoventilation.

Non-invasive ventilation (i.e. without an endotracheal or tracheostomy tube) with *bilevel positive-pressure ventilation* provides an inspiratory pressure support (ventilation) and expiratory pressure support to prevent respiratory collapse. The difference between the inspiratory and expiratory pressure enhances the tidal volume, reducing the work of respiratory muscles. It consists of a ventilator connected to the nasal mask without the need for intubation. This device allows other physiological functions such as speaking and eating. Non-invasive positive pressure ventilation improves quality of life and survival. Non-invasive positive pressure is used to treat hypoventilation related to chronic obstructive pulmonary disease, Obesity Hypoventilation Syndrome (OHS), neuromuscular disorders, chest deformity, and idiopathic disease (**Fig. 4-6**).

Volume assured pressure support delivers a patient-triggered breath utilizing a variable pressure support, to allow an adequate tidal volume or minute ventilation during sleep. It is mainly used in hypoventilation associated with neuromuscular disorders, primary lung disease, OHS, COPD, or chest deformities, especially when bilevel fails.

Figure 4-7 Summarizes the indications for the different positive airway pressure devices.

FIGURE 4-5 TREATMENT OF HYPOVENTILATION

- Correct the underlying disorder.
- Medications
  - — Bronchodilators
  - — Respiratory stimulants: medroxyprogesterone, acetazolamide and theophylline
  - — Naloxone to reverse the effect of narcotics (when indicated)
  - — Flumazenil to reverse the effect of benzodiazepines (when indicated)
- Weight loss
- Supplemental oxygen
- Non-invasive ventilation with bilevel positive-pressure ventilation
- Volume pressure support

FIGURE 4-6 INDICATIONS FOR NOCTURNAL BILEVEL POSITIVE-PRESSURE VENTILATION

- Symptoms and signs of nocturnal hypoventilation
- Shortness of breath at rest
- Failure to respond to first-line therapy measures
- Moderate to severe hypoventilation
- Pulmonary hypertension and cor pulmonale
- Carbon dioxide >45 mmHg
- Nocturnal hypoxia (arterial oxygen saturation < 88%)

FIGURE 4-7 INDICATIONS FOR POSITIVE AIRWAY PRESSURE DEVICES	
Continuous positive airway pressure (CPAP)	• First-line therapy for moderate to severe obstructive sleep apnea • Mild obstructive sleep apnea associated with symptoms or comorbid disorders • Cheyne-Stokes breathing
Bilevel positive airway pressure (BiPAP)	• Patients with obstructive sleep apnea who cannot exhale against the high pressure of the CPAP • Obstructive sleep apnea with persistent oxygen desaturation due to hypoventilation despite CPAP • Obstructive sleep apnea and chronic hypercapnic respiratory failure (such as obesity hypoventilation syndrome, neuromuscular disease, brain stem disease) whose alveolar hypoventilation failed to respond to CPAP therapy • Patients with *overlap syndrome* (obstructive sleep apnea and chronic obstructive pulmonary disease) • Some patients with obstructive sleep apnea and central sleep apnea • Cheyne-Stokes breathing that failed CPAP and adaptive servo-ventilation (ASV) • Idiopathic central sleep apnea
Auto-positive airway pressure (APAP)	• Patients with moderate to severe obstructive sleep apnea with daily sleepiness who did not tolerate CPAP
Auto-bilevel positive airway pressure	• Patients with obstructive sleep apnea with poor initial experience to CPAP
Adaptive servo-ventilation (ASV)	• Central sleep apneas coexistent with obstructive sleep apnea • Cheyne-Stokes breathing that failed CPAP • Complex sleep apnea • Central sleep apnea related to narcotic usage
Non-invasive positive pressure ventilation	• Patients with persistent sleep-related hypoventilation and carbon dioxide retention despite CPAP therapy

5

Insomnia

Insomnia is the complaint, for at least a month, of difficulty falling asleep, difficulty staying sleep, early-morning awakening, or non-refreshing/non-restorative sleep despite an adequate opportunity to sleep. Insomnia is associated with impaired daytime functions, including fatigue, impaired cognitive function, irritability, poor multitasking skills, decreased work or school performance, depressed mood; and proclivity to accidents, falls, and pain. In children, caretakers report insomnia as bedtime resistance or inability to sleep independently, or both.

Demographics of Insomnia

Insomnia is the most common sleep disorder in clinical practice. Its prevalence is higher among women, older adults, and people with psychiatric or medical disorders. One in three adults report occasional insomnia symptoms, and one in ten adults suffers from chronic insomnia. A positive family history of insomnia, lower socioeconomic status, unemployment, marital status problems (divorced or widowed), and shift work are risk factors that predispose individuals to insomnia.

Pathophysiology of Insomnia

Insomnia may result from the interplay of three essential factors, also referred to as *Spielman's 3Ps* (**Fig. 5-1**).

The Biology of Insomnia

Abnormalities of the circadian sleep/awake rhythm and homeostasis contribute to the genesis of insomnia. *Hyperarousals* and *increased cortical arousals* are cardinal features.

Hyperarousals are an increase in physiological parameters (heart rate, temperature, metabolic rate, and cortisol levels) caused by an increment in the ascending reticular activating system or decrement in the sleep drive.

Increased cortical arousals are electroencephalogram changes, such as an increment in wakefulness waves

FIGURE 5-1 FACTORS CAUSING AND MAINTAINING INSOMNIA	
Predisposing factors	Female gender, genetics, anxiety, personality, early life trauma, lack of enforcement of routine sleep hours during childhood, poor relationships
Precipitating factors	Acute illness, job change or loss, separation, assault, death of a loved one, other stressful event
Perpetuating factors	Excessive worry, spending too much time in bed trying to sleep, using the bed for working or activities other than sleep or sex, naps, fear of sleeplessness

(alpha and beta) and decrement in deep sleep waves (theta and delta), compared with those of normal sleepers.

Both psychological and biological factors contribute to the development of insomnia.

Classification of Insomnia

Insomnia can be divided into *primary* and *secondary insomnia*.

Primary insomnia, also called *idiopathic insomnia,* is not associated with any other underlying medical, neurological, or psychiatric disorder, or with medication/substance use, abuse or withdrawal.

Secondary insomnia, also called *comorbid insomnia,* is associated with other medical, neurological, psychiatric disorders, or with medication/substance use, abuse or withdrawal.

Primary sleep disorders may cause a secondary insomnia:

- *Restless leg syndrome* and delayed circadian rhythm disorders may cause *sleep-onset insomnia* (problems initiating sleep).
- *Periodic leg movement disorder,* obstructive sleep apnea, central sleep apnea, parasomnias, and narcolepsy may cause *sleep-maintenance insomnia* (problems staying asleep). Parasomnias are undesirable physical and behavioral phenomenon during sleep **(Chapter 8).**
- Advanced circadian rhythm disorders may cause *early awakening*. These sleep disorders will be discussed in more detail in separate chapters. **Figure 5-2** shows the International Classification of Insomnias.

Except for adjustment insomnia (a form of acute insomnia), which lasts less than three months, the other types of insomnia discussed below are chronic (lasting more than three months). 10-15% of patients with chronic insomnia have a primary insomnia, while 80-90% have a secondary insomnia.

FIGURE 5-2 INTERNATIONAL CLASSIFICATION OF INSOMNIAS

PRIMARY INSOMNIAS	**SECONDARY INSOMNIAS**
Adjustment Insomnia	Insomnia Due to Psychiatric Disorder
Psychophysiological Insomnia	Insomnia Due to Neurological Disorder
Paradoxical Insomnia	Insomnia Due to Medical Disorder
Idiopathic Insomnia	Insomnia Due to Drug or Substance
Inadequate Sleep Hygiene	Insomnia Due to Other Sleep Disorders
	Behavioral Insomnia of Childhood

Adjustment insomnia is triggered by an identifiable stressor and lasts a few days to a few weeks, but less than three months. This type of insomnia resolves when the stressor is no longer present or when the patient adapts to it. For example, an old lady developed a six-week insomnia soon after the sudden death of her husband.

Psychophysiological insomnia results from heightened arousal or learned sleep-preventing associations at bedtime. These patients have a state of cognitive *hypervigilance*, or "racing mind" when they have to go to sleep (**Fig. 5-3**). For example, a college student presents with six-month history of insomnia; at bedtime she was preoccupied with not being able to fall asleep; she constantly looked at the clock and finally she was so agitated she could not fall asleep.

Paradoxical insomnia. Patients complain of chronic severe insomnia (very minimal or no sleep) but with disproportionally less objective evidence of daytime impairment. Patients overestimate the time to fall sleep and underestimate the total sleep time. For example, once a lady came to see me with a six-month history of not being able to sleep at all. However, I noticed minimal objective signs of sleep deprivation. She never took naps during the daytime, and she seemed more alert than most post on-call physicians I know.

Idiopathic insomnia is a long-lasting insomnia that starts in childhood and continues chronically during adulthood, not associated with any identifiable cause.

Inadequate sleep hygiene insomnia results from voluntarily engaging in practices that increase arousal and disrupt the propensity to fall asleep. Examples are working late at night, having irregular sleep hours, taking long or late afternoon naps, high coffee intake,

Figure 5-3. *Hypervigilance before sleep.*

watching TV at bedtime, and trying to sleep with the lights on.

Insomnia due to a psychiatric disorder and *insomnia due to a medical/neurological disorder* are forms of insomnia caused by psychiatric and medical conditions, respectively. The insomnia starts with the disease onset and waxes and wanes in unison with the other symptoms of the disease. The most common psychiatric disorders causing insomnia are depression and anxiety. Medical disorders associated with insomnia include chronic painful disorders (fibromyalgia, neuropathic pain, back pain), obstructive pulmonary disease (emphysema, asthma), cardiac disorders (congestive heart failure), and neurological disorders (dementia, Parkinson's disease). **Figure 5-4** lists different psychiatric and medical conditions associated with problems of falling asleep, staying asleep, or causing early morning awakening.

Delayed sleep phase syndrome: sleep-onset and wake time are > 2 hrs. delayed in relationship to conventional sleep/wake time.

Advanced sleep phase syndrome: sleep-onset and wake time are > 2 hrs. earlier than conventional sleep/wake time.

Insomnia due to medication or substance abuse resembles insomnia due to a medical or psychiatric disorder, but is instead caused by a substance or medication that is being used, abused, or withdrawn (**Fig. 5-5**). Alcohol, caffeine and nicotine are the most commonly abused substances.

FIGURE 5-4 SECONDARY CAUSES OF INSOMNIA
SLEEP ONSET INSOMNIA
Anxiety
Depression
Acute psychosis
Delayed sleep phase syndrome (**Chapter 7**)
Restless leg syndrome
Substances (coffee, decongestants)
Chronic pain syndromes
Sleep disorder of breathing
SLEEP MAINTENANCE INSOMNIA
Major depression
Sleep disorder of breathing
Periodic leg movements of sleep
Chronic pain
Respiratory or cardiac disorders
Neurological disorders (Parkinson's disease, multiple sclerosis)
Nocturia
Poor sleep environment
EARLY MORNING AWAKENING
Major depression
Awakening early to fulfill family or working responsibilities
Drugs
Advanced sleep phase syndrome (**Chapter 7**)

FIGURE 5-5 MEDICATIONS ASSOCIATED WITH INSOMNIA
Antidepressants
Bronchodilators
Corticosteroids
Thyroid hormones
Decongestants
Anorexiants (appetite suppressants)
Diuretics
Beta blockers
Dopamine agonists and antagonists
Narcotic analgesics
Stimulants
Anti-epileptics

Insomnia due to other sleep disorders is discussed above.

Behavioral insomnia of childhood is a form of childhood insomnia with problems falling asleep, staying asleep, or both, with a behavioral etiology. There are two types: *limit-setting type,* where resistance to going to sleep results from inadequate enforcement of bedtime from the caregiver; and *sleep-onset association type,* where there is a problematic precondition required to fall asleep. For example, a three-year-old boy cannot fall asleep unless he watches TV in his bedroom, has a bottle or a pacifier, or sleeps with his parents.

Evaluation of Insomnia

The evaluation of insomnia starts with a pertinent history and physical exam:

- Two-week sleep diaries may help in the diagnosis of insomnia.
- Are there medical or psychiatric disorders?
- Is there a drug abuse?
- If you suspect sleep apnea, check for abnormalities of the head and neck (face; neck diameter; upper airway obstruction; etc.).
- Do a neurologic exam if restless leg or other neurological syndromes are suspected.
- Examine the chest and legs if a cardiovascular and/or respiratory condition is implicated.
- Do laboratory tests to assess comorbid conditions.
- Do nocturnal polysomnography for suspected obstructive sleep apnea, and actigraphy for suspected periodic leg movement disorder, circadian rhythm disorder, or paroxysmal insomnia.

Treatment of Insomnia

The treatment of insomnia aims to correct night sleep disturbances and daytime impairment. There are two forms of therapy for insomnia: *non-pharmacologic* and *pharmacologic*.

Non-pharmacologic treatment. Cognitive behavioral therapy focuses on gaining control over sleep by modifying sleep/awake schedules, restricting time spent in bed when the patient is fully awake, and avoiding the bed for anything other than sex or sleep. Other helpful interventions include learning different relaxation techniques, changing maladaptive behaviors that interfere with sleep, focusing on dysfunctional beliefs, and decreasing worries and fears that decrease proclivity to fall asleep ("If I do not sleep, I cannot function"; "If I do not sleep, I will die"). Implementing proper sleep hygiene techniques and sustaining a regular morning wake-up time are essential in treating insomnia as well as other sleep disorders (**Fig. 5-6**).

Cognitive behavioral therapy is effective in chronic insomnia and is considered the first-line therapy for primary insomnia and insomnia caused by chronic pain, cancer, or depression. Cognitive behavioral therapy provides long-term benefit beyond the duration of the therapy, contrary to pharmacotherapy, which loses its benefit after drug discontinuation. The drawback of cognitive behavioral therapy is the limited number of trained providers available to administer biweekly sessions, and the cost. Cognitive behavioral therapy can be offered as an internet-based video conference for people unable to access it in person.

Additional successful techniques used in insomnia are mindfulness meditation and acupuncture. Acupuncture increases total sleep time probably because it increases melatonin secretion and reduces stress and anxiety. It has also been utilized in post-stroke insomnia and post menopause insomnia. Acupuncture alone or in combination with a sedative benzodiazepine improves sleep latency and total sleep time. Hypnosis might also help with insomnia, but it has the potential drawback of resurfaced stressful memories. All these techniques require further investigation.

Yoga, tai-chi, and mindfulness meditation are used to treat insomnia. Mindfulness meditation reduces stress and improves sleep in cancer patients.

Alpha cranial electrical stimulation is an FDA-approved hand-held non-invasive device that delivers alternating microcurrents via small clips applied to the ear lobe or other scalp regions. It decreases the mean alpha wave frequency (relaxing effect), as seen in mindful meditation.

Low frequency repetitive transmagnetic stimulation of the brain is a non-invasive technique that improves sleep quality in chronic insomnia, probably by inhibiting hyper-arousal symptoms.

Open-loop audio-visual stimulation is a form of neurofeedback program that is self-administered for 30 min at night at home as light-sound patterns to induce slow wave sleep.

Other devices used in insomnia are weight blankets, which increase total sleep time; cool mattresses, pillow, and blankets used in patients with insomnia related to heat-intolerance, as in postmenopausal women and patients with multiple sclerosis.

Pharmacological therapy. Different drug groups, both prescription and non-prescription hypnotics (drugs that enhance drowsiness), have been used to treat insomnia. These include benzodiazepines, non-benzodiazepine receptor agonists, antidepressants, and others. Some hypnotics have a short duration of action (half-life less than 3 hrs.) and are mainly used in sleep-onset insomnia. Other drugs have an intermediate duration of action (half-life > 4 hrs.); these are mainly used in sleep-maintenance insomnia (**Fig. 5-7**).

Benzodiazepines (*triazolam, temazepam, estazolam*) bind to gamma-aminobutyric acid (GABA) benzodiazepine receptor complex. Benzodiazepines bind to different subunits of this complex, having effects besides hypnotic, such as anticonvulsant, myorelaxant, and anxiolytic. Benzodiazepines must be used cautiously in chronic insomnia due to the risk of abuse, tolerance and withdrawal, as well as ataxia, falls, cognition and psychomotor impairment, rebound insomnia, overdose, and aggravation of obstructive sleep apnea.

Benzodiazepine agonists (*zolpidem, zaleplon, eszopiclone*) are FDA-approved drugs for insomnia with less danger of abuse, tolerance, and withdrawal symptoms than benzodiazepines, due to their selective effect only on the GABA benzodiazepine receptor complex type-1. They have less risk of rebound insomnia, cognition impairment, and fatal overdose than benzodiazepines.

FIGURE 5-6 SLEEP HYGIENE TECHNIQUES
Get enough sleep per night. Most people need between 7 and 8 hours of sleep a night, no less than 6.5 hours.
Avoid naps if possible. If naps are necessary, prefer scheduled naps late in the morning or early in the afternoon.
Maintain a regular sleep schedule 7 day of the week. Get up at the same time each morning regardless the time you go to bed.
Avoid alcohol, especially in the evenings.
Avoid caffeine after lunch.
Avoid nicotine in the evenings and at night.
Avoid exercise before dinnertime.
The sleep environment needs to be suitable for sleep.
Reserve the bed for sleeping or sex only.
Sleep when you are sleepy; otherwise, get out of bed after 15 to 20 minutes if you are not sleepy, and engage in some sedentary activity until you begin to feel sleepy; then go back to bed.

FIGURE 5-7 PHARMACOLOGICAL THERAPY OF INSOMNIA				
OS = oral spray; PO = per ora SL = sublingual; SLL = sublingual lozenge				
Akathisia: movement disorder that results from a feeling of constant inner restlessness. Paresthesias: abnormal skin sensations.				
	Brand name	**Dosage**	**Frequent side effects**	**Other clinical uses**
Short Duration of Action				
Zolpidem	Ambien	PO: 5-10mg OS: 5-10mg SL: 5-10mg SLL: 1.75-3.5mg	Headaches, dizziness, somnolence	
Zaleplon	Sonata	5-20mg	Dizziness, lightheadedness, headaches, paresthesias, nausea, abdominal pain, somnolence	
Ramelteon	Rozerem	8mg	Drowsiness, fatigue, dizziness; might affect liver function	Insomnia comorbid with COPD, sleep breathing disorder
Triazolam	Halcion	0.125-0.25mg	Rebound insomnia and anxiety, amnesia, confusion	
Valerian		400-900mg		
Intermediate Duration of Action				
Temazepam	Restoril	7.5-30mg	Daytime drowsiness, cognitive impairment, impaired motor coordination, falls, dizziness, respiratory depression	Insomnia comorbid with anxiety
Estazolam	Prosom	1mg-2mg	Daytime drowsiness, cognitive impairment, impaired motor coordination, falls, respiratory depression	Insomnia comorbid with anxiety
Eszopiclone	Lunesta	1-3mg	Metallic taste, headaches, dizziness, somnolence, next-day impairment in some patients	
Zolpiden-extended release		6.25-12.5mg (6.25mg is the max dose in women)	Headaches, dizziness, somnolence, next-day impairment in some patients	
Trazodone	Desyrel	25-200mg	Sedation, dizziness, headaches, orthostatic hypotension, blurry vision, priapism, amnesia	Insomnia comorbid with major depression Insomnia comorbid with substance abuse
Amitriptyline	Elavil	10-100mg	Sedation, weight gain, dry mouth, constipation, orthostatic hypotension, urinary retention	Insomnia comorbid with depression, anxiety, and pain
Doxepin	Sinequan	3-6mg	Sedation, nausea and upper respiratory infection. At higher dose (>6mg) it may cause weight gain, dry mouth, constipation, orthostatic hypotension, urinary retention.	Insomnia comorbid with depression, anxiety, and pain

(Continued)

Mirtazapine	Remeron	7.5-30mg	Weight gain, increased appetite, dry mouth, constipation	Insomnia comorbid with depression Insomnia comorbid with sleep breathing disorders
Suvorexant	Belsomra	5-20mg	Sedation	Early morning awakening with difficulty resuming sleep Mild to moderate obstructive sleep apnea
Quetiapine	Seroquel	25-200mg	Sedation, orthostatic hypotension, dry mouth, tachycardia, weight gain	Insomnia comorbid with psychosis
Olanzapine	Zyprexa	2.5-20mg	Sedation, agitation, dizziness, orthostatic hypotension, akathisia, weight gain. Increased incidence of cerebrovascular events in demented patients	Insomnia comorbid with psychosis
Gabapentin	Neurontin	100-900mg	Dizziness, drowsiness, weight gain	Insomnia comorbid with seizures, pain, restless leg syndrome
Pregabalin	Lyrica	75-300mg	Dizziness, drowsiness, weight gain, cognitive impairment, increased appetite	Insomnia comorbid with seizures, fibromyalgia, pain, or restless leg syndrome
Diphenhydramine	Benadryl	25-50mg	Sedation, dizziness, dry mouth, blurry vision, constipation, urinary retention, cognitive impairment	Insomnia comorbid with allergy or upper respiratory infection
Doxylamine	Unisom	25-50mg	Sedation, dizziness, dry mouth, blurry vision, constipation, urinary retention, cognitive impairment	Insomnia comorbid with allergy or upper respiratory infection
Suvorexant	Belsomra	5-20mg	Sedation, headaches	Insomnia comorbid with Alzheimer
Daridorexant	Belsomra	25-50mg	Nasopharyngitis, headaches	
Lemborexant	Dayvigo	5-10mg	Somnolence, nightmares and potential attention and memory impairment	

The *melatonin-receptor agonist ramelteon* is an FDA-approved medication for insomnia with no known abuse potential that is contraindicated in severe liver disease. Regular *melatonin* is commonly used as an over-the-counter drug; however, its main benefit has been on insomnia related to circadian rhythm disorders, rather than general insomnia.

Sedative antidepressants such as *doxepin*, *amitriptyline* and *trazodone* have been used in sleep-onset and sleep-maintenance insomnia, especially with comorbid depression, anxiety, and pain. However, antidepressants have potential anticholinergic side effects — cardiac arrhythmia, orthostatic hypotension, urinary retention, weight gain, blurry vision, daytime sedation, and REM suppression that results in REM sleep rebound with disturbed dreams, if acutely discontinued. *Trazodone* has fewer anticholinergic side effects than tricyclic antidepressants (doxepin and amitriptyline), but it may cause priapism (prolonged, painful erection). *The ultra-low dose of doxepin* (3-6mg) is FDA approved to treat sleep-maintenance insomnia via a selective H1-receptor antagonist effect. However, higher doses of doxepin up to 50 mg at night have also been used, but carry a higher toxicity risk. *Mirtazapine*, another hypnotic antidepressant, is unfortunately associated with weight gain. Trazodone and mirtazapine have low abuse potential, so they may be used in individuals with substance abuse disorders.

Orexin-receptor antagonists are a newer group of FDA-approved drugs to treat sleep-onset and sleep-maintenance insomnia. These drugs reduce sleep latency, increase total sleep times, and reduce wake after sleep onset. It includes three drugs: *suvorexant*, *lemborexant*, and *daridorexant*. These drugs are not

FIGURE 5-8 OVER-THE-COUNTER SLEEP AIDS FOR INSOMNIA

NAME	MECHANISM OF ACTION	DOSE QHS	SIDE EFFECTS
Melatonin	Reduces sleep latency; improvement in sleep quality	1-10 mg	Nightmares, agitation, anxiety, and drowsiness
L-tryptophan	Improves quality of sleep, reduces wake-after-sleep onset	250 mg	No serious adverse effect
Valerian-Hops	Reduces sleep latency	374 mg native extract and hop 83.8 native extract	No serious adverse effect. Contaminants may cause hepatotoxicity
Diphenhydramine	Reduces sleep latency, improves quality of sleep, reduces wake-after-sleep onset, increases total sleep time	25-50mg	Sedation, dizziness, dry mouth, blurry vision, constipation, urinary retention, cognitive impairment
Doxylamine	Reduces sleep latency	25-50mg	Sedation, dizziness, dry mouth, blurry vision, constipation, urinary retention, cognitive impairment
Ginseng	Increases sleep time and increases slow wave	40 mg ginger extract	Nervousness and excitation
Kava Kava	Decreases sleep latency, decreases wake after sleep onset, increases sleep time	70-250mg kavalactone, which is equivalent to at least 300mg kava root extract	Not known
Chamomile	Minor sedative	Tea: 1 cup Tincture: 30 drops	Not known
Passion Flower	Minor sedative	4-8 grams as tea	Uterine stimulant, so contraindicated in pregnancy
St. John Wort	Increases slow wave sleep (deep sleep)	Capsules 300mg; tablets 450mg; tea one cup or ½ to 1 teaspoon	Sedation, dry mouth, dizziness, gastrointestinal upset, restlessness, hypersensitivity
Semen Ziziphi	Increases total sleep time and slow wave sleep (deep sleep)	24-36 grams	Gastrointestinal symptoms, dizziness, rash
Wuling	Increases sleep time	One to three capsules (0.33 grams each)	Headaches, dizziness, gastrointestinal symptoms, limb numbness, dry mouth

associated with rebound insomnia and/or withdrawal effects the following morning. However, suvorexant and leborexant may be associated with next morning driving impairment and individual should be cautioned about this side effect especially when using the highest doses. Suvorexant received approval also to treat insomnia related to Alzheimer dementia . Suvorexant may cause abnormal dreams, inducing parasomnias (sleepwalking or eating), and may be habit-forming. These drugs are not recommended in narcolepsy (**Chapter 6**).

Other prescription or non-prescription drugs used in insomnia are sedative antipsychotics (*quetiapine, olanzapine*), antihistamines (*diphenhydramine, doxylamine*), antiepileptics (*gabapentin, pregablin*), and botanic preparations (*kava* and *valerian*), although the benefit of the botanic preparations remains unproven.

Antipsychotic sedatives have antagonistic effects on multiple types of brain receptors, including those involving dopamine and histamine. Antipsychotics have the potential to cause parkinsonism and *tardive dyskinesia* (involuntary movements of the face and jaw; anti-dopamine effect). *Olanzapine* has a long latency onset so is only used in sleep-maintenance insomnia. *Quetiapine* may increase the Q-T interval on the electrocardiogram and should be used cautiously in cardiac disease.

Gabapentin and *pregabalin* have hypnotic effects, especially helpful in coexistent restless leg syndrome, periodic leg movement during sleep, seizures, and pain; the dose needs to be adjusted to renal function.

Diphenhydramine and *doxylamine* are commonly found in over-the-counter medications, but their benefits need to be counter-balanced with daytime sleepiness, rapid-onset tolerance, and anti-cholinergic effects such as cognitive impairment, urinary retention, and constipation.

Alcohol is probably the most common substance used to fall asleep. Misinformed individuals consume it because it reduces sleep latency; however, its fast metabolism causes increased arousal after sleep onset in the second half of the night.

Over-the-counter herbal products are routinely used as sleeping aids for insomnia. Herbs do not carry, as prescribed drugs do, any significant risk of overdose, tolerance, habituation, or addiction. However, their benefit and safety remains unknown. **Figure 5-8** shows a list of over-the-counter herbs used as sleeping aids in insomnia, but not yet recommended by the American Academy of Sleep Medicine.

Melatonin is used in insomnia and especially in delayed sleep/wake phase syndrome, jet lag, and night shift work. This is further discussed in the chapter on circadian sleep/wake rhythm disorders **(Chapter 7)**. Melatonin is not associated with rebound and withdrawal effects.

6

Narcolepsy and Idiopathic Hypersomnia

Narcolepsy and *idiopathic hypersomnia* are two neurologic disorders that present with excessive daytime sleepiness not attributed to disturbed or insufficient nocturnal sleep, sleep breathing disorders, or circadian rhythm disorders.

Narcolepsy

Narcolepsy is characterized by a tetrad of

- excessive daytime sleepiness
- cataplexy
- sleep paralysis
- sleep hallucinations

Excessive daytime sleepiness is the inability to stay awake and alert during the day. It is the most disabling of all the symptoms and is present in almost all narcolepsy patients to different degrees. It presents with daily drowsiness and frequent daily naps, especially during boring activities. Most patients feel refreshed when awakening, but drowsiness resumes minutes or hours later. Irresistible episodes of sleepiness can occur suddenly without warning.

Cataplexy consists of transient muscle weakness triggered by an intense emotion, usually laughing. Weakness may be focal, presenting with eyelid drooping, mouth sagging, head nodding, or knee buckling. It may also be generalized, with patients falling to the ground. Respiratory and extra-ocular muscles are spared. Cataplexy lasts from seconds to about two minutes, with immediate and full recovery. Consciousness and memory are preserved in cataplexy, in contrast to epilepsy where consciousness and memory are lost. Cataplexy presents in 60% of narcoleptics, often within three to five years of the onset of excessive daytime sleepiness.

Sleep paralysis differs from cataplexy in that it occurs upon falling asleep or awakening, while cataplexy occurs while the patient is already up and is triggered by emotions. The patient feels awake in sleep paralysis, but cannot move. It occurs more commonly when sleeping supine and may last seconds to minutes.

Sleep hallucinations are called *hypnagogic* when they occur while transitioning from wakefulness to sleep. They are called *hypnopompic* when they strike while waking up. People with sleep hallucinations hear voices, see strange people or objects in the room, or have other strange sensations.

Narcolepsy is divided into *narcolepsy Type 1* (narcolepsy with cataplexy) and *narcolepsy Type 2* (narcolepsy without cataplexy). Narcolepsy with cataplexy is the most common form and the one discussed here. Narcolepsy Type 2 is a poorly understood non-specific heterogeneous sleepiness disorder, but its real prevalence is unknown. According to some data it might be more common than narcolepsy Type 1.

Narcolepsy starts during adolescence and young adulthood, affecting both sexes equally. Narcoleptics may have sleep fragmentation, memory lapses, and automatic behavior (a semiconscious state where patients engage in activities while still sleepy, without any recollection). Narcolepsy affects about 0.05% of the population of the United States and Europe, being much more common in Japan and much less common in Israel.

Pathophysiology of Narcolepsy

Narcolepsy with cataplexy results from loss of hypocretin-containing neurons in the lateral hypothalamus. *Hypocretin* (*orexin*) is a neuropeptide that promotes wakefulness and prevents inappropriate transitions between being awake and asleep. Narcolepsy is probably an autoimmune process with a genetic background based on the high prevalence of *human leukocyte antigen DQB1*0602* in narcoleptics, especially patients with low hypocretin levels in their spinal fluid (found in about 98% of patients with narcolepsy with cataplexy). Hypocretins normally increase wakefulness and reduce rapid-eye-movement (REM) sleep. Reduction of hypocretin may be responsible for many of the features of narcolepsy.

Diagnosis of Narcolepsy

The diagnosis of narcolepsy with cataplexy is made on history alone when patients present with the classic tetrad. However, only a third of patients present with the whole tetrad. Otherwise, the diagnosis depends on clinical suspicion and one of the following: abnormal nocturnal polysomnography and daytime multiple sleep latency test, *or* low hypocretin levels in the spinal fluid.

In narcolepsy, *nocturnal polysomnography* displays spontaneous awakenings, reduced sleep efficiency (percentage of the time spent in bed sleeping), and REM sleep within 20 minutes from sleep onset (healthy individuals display REM sleep onset after 80 to 100 minutes into sleep). Nocturnal polysomnography may reveal other sleep disorders that may coexist with narcolepsy and contribute to excessive daytime sleepiness. The *Multiple Sleep Latency Test* shows a short *sleep onset latency* (time from lights out to first sleep epoch), less than 8 minutes, and two or more naps with *sleep onset REM*. Most patients with narcolepsy have spinal fluid hypocretin levels of 110 pg/ml or less (30% of the mean control value). Human leukocyte antigen DQB1*0602 is not usually tested in the diagnosis of narcolepsy, since it is often seen in normal individuals. DQB1*0602 has more value when negative than when positive since it makes the diagnosis of narcolepsy unlikely and also unlikely to have low orexin levels in the spinal fluid, which goes against performing this procedure.

Figure 6-1 displays the approach to evaluating excessive daytime sleepiness.

Treatment of Narcolepsy

Treatment aims to enhance alertness and reduce or eliminate cataplexy, hallucinations, and sleep paralysis. Other sleep disorders that might coexist with narcolepsy, such as sleep apnea, periodic leg movements during sleep, and REM-behavior disorder also need to be treated.

There are two forms of therapy for narcolepsy: non-pharmacological (**Fig. 6-2**) and pharmacological.

Figure 6-2 summarizes the non-pharmacologic treatment of excessive daytime sleepiness.

Pharmacological Therapy of Excessive Daytime Sleepiness (EDS)

Three groups of drugs are commonly used to treat Excessive Daytime Sleepiness (EDS): non-amphetamine wake-promoting agents, amphetamine wake-promoting agents, and *sodium oxybate*.

Non-amphetamine wake-promoting agents are the first line of therapy for narcolepsy. There are two drugs in this group: *modafinil* (*Provigil*) and *armodafinil* (*Nuvigil*). *Modafinil* is usually initiated at a dose of 200 mg each morning and often titrated up to 300 or 400 mg in two divided doses, last dose no later than noon. *Armodafinil* is give once daily in the morning starting at 150 mg and increased to 250 mg if necessary. These drugs enhance dopaminergic and aminergic transmission in the anterior hypothalamus. They have no issues with tolerance (a need for a higher dose over time to sustain the same benefit) or withdrawal symptoms. They have lower risk of abuse than amphetamines. Headaches and nausea are common side effects.

Amphetamine-like wake-promoting agents, also known as stimulants, including *methylphenidate* and *amphetamines*.

Methylphenidate is a dopamine reuptake inhibitor with an amphetamine-like action. It is better tolerated than amphetamines, with less risk of abuse and fewer side effects. It is considered a second-line therapy for Excessive Daytime Sleepiness (EDS) since it has fewer amphetamine-like effects, discussed below. Methylphenidate comes in short- and long-acting preparations. The initial dose is 10 mg twice daily, the last dose given no later than 3 pm to prevent insomnia. Methylphenidate usually is used at 10-60 mg in two or three divided doses, *or* once daily with extended-release preparations.

Amphetamines enhance dopamine and norepinephrine release and inhibit their reuptake. Amphetamines treat Excessive Daytime Sleepiness, cataplexy, and other REM-related phenomena. Amphetamines improve alertness, attention, vigilance, memory, and performance. The most used amphetamines are *dextroamphetamine*,

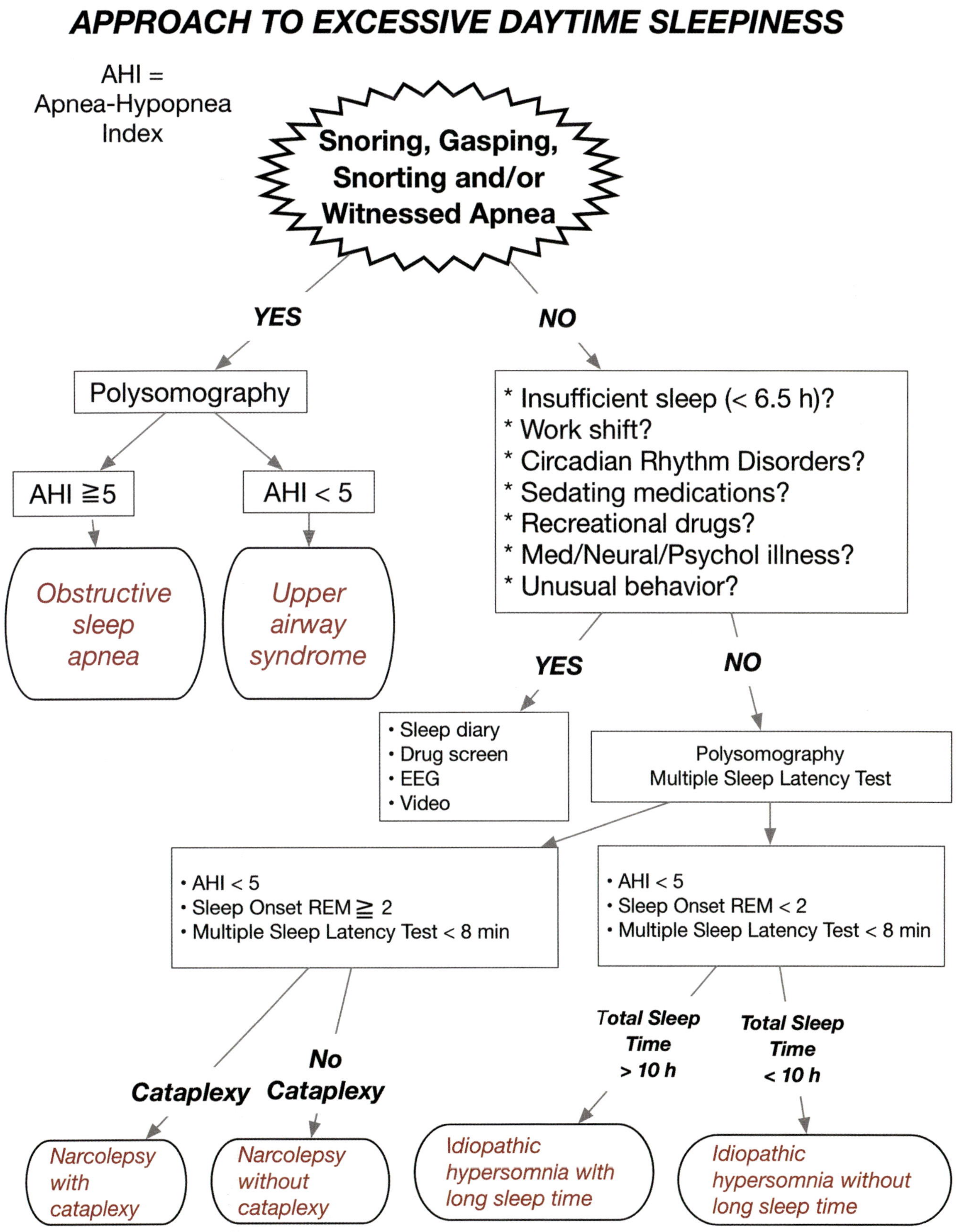

Figure 6-1. *Evaluation of excessive daytime sleepiness.*

FIGURE 6-2 NON-PHARMACOLOGICAL TREATMENT OF EXCESSIVE DAYTIME SLEEPINESS

- Avoid sleep-inducing medications, e.g. benzodiazepines, antipsychotics, opioids
- Avoid drugs that cause insomnia, e.g. caffeine in the afternoon and/or evening
- Avoid alcohol, which aggravates Excessive Daytime Sleepiness and Obstructive Sleep Apnea
- Keep regular sleep-awake schedules
- Schedule 20-minute naps once or twice daily
- Implement sleep hygiene techniques **(Fig. 5-6)**
- Avoid sleep deprivation
- Psychosocial support (e.g. eliminate misconceptions about sleep; attend group therapy; prevent obesity)

methamphetamine, and their combination. The starting dose is 5 mg twice daily, with the last dose no later than 3 pm. Therapy is usually initiated with an immediate-release preparation and later changed to a sustained-release formulation when an effective dose is established. They are considered third-line therapy (after methylphenidate) because of their adverse effects of nervousness, insomnia, anorexia, anxiety, headaches, hypertension, palpitation, depression, psychosis, paranoia, postural hypotension, Q-T prolongation, and ventricular arrhythmias. Amphetamines have a high risk of abuse and tolerance. Therefore, they are highly regulated. Amphetamines and methylphenidate should not be used in patients with cardiovascular disease or patients taking MAO-inhibitors, due to the risk of inducing severe hypertension. **Figure 6-3** shows the basic principles of using wake-promoting drugs.

Sodium oxybate (Xyrem) is effective and FDA-indicated for excessive daytime sleepiness and cataplexy. It can be used as first-line therapy when both conditions are distressing enough (**Fig. 6-4**). It improves sleep efficiency (time in bed sleeping). Its mechanism of action remains unknown, but it works on GABA-B receptors. It decreases sleep fragmentation and increases slow-wave sleep. It is given as a liquid at bedtime with a second dose 2.5 to 4 hours later due to its short half-life. The starting dose is 3g twice per night; the dose can be increased over weeks to 4.5g twice nightly. Adverse effects include headaches, nausea, dizziness, enuresis, deep sedation, and aggravation of sleepwalking. This drug is highly regulated because it is misused as a "rape drug," being a colorless and tasteless liquid that may be added to a drink to induce sleepiness.

Solriamfetol is a newer selective low potency dopamine and norepinephrine reuptake inhibitor with wake-promoting effects, used for excessive daytime sleepiness. The dosage is 150-300mg day. Most common side effects are headaches, nausea, and loss of appetite.

FIGURE 6-3 GENERAL PRINCIPLES OF DRUG USE FOR EXCESSIVE DAYTIME SLEEPINESS

- Drugs should be taken upon awakening and no later than noon to prevent insomnia; the exception is sodium oxybate, which is taken at night.
- Drugs should be started at the lowest possible effective dose.
- Drugs require close monitoring for potential side effects.
- Drugs should not be used to replace insufficient sleep.

Serious side effects are tachycardia, hypertension, and psychiatric symptoms.

Pitosilant is a H3-histamine receptor antagonist/inverse agonist used to treat excessive daytime sleepiness and cataplexy. The dosage ranges from 5 to 40mg. Its most common side effects are headache, anxiety, nausea, and irritability. Serious side effects are prolongation of the QT interval.

REM-intrusion phenomena include cataplexy, hallucinations, and sleep paralysis, but only cataplexy is disabling enough to require therapy. Drugs that increase noradrenaline and serotonin inhibit the generation of REM and therefore prevent REM-intrusion phenomena. Most drugs for cataplexy are antidepressants that increase serotonin and norepinephrine levels, such as *venlafaxine*, *protriptyline*, and *clomipramine*. Selective serotonin reuptake inhibitor antidepressants only increase the serotonin level, and are thereby less effective in cataplexy. Abrupt discontinuation of antidepressants may trigger *status cataplecticus*, a state of continuous cataplexy that lasts for hours to days. *Sodium oxybate* is effective for cataplexy, but is reserved for refractory cases.

Idiopathic Hypersomnia

Idiopathic hypersomnia is a neurological disorder that, like narcolepsy, presents with Excessive Daytime Sleepiness (EDS) despite sufficient sleep time. Nighttime sleep is non-refreshing, and patients suffer from sleep inertia — persistent grogginess after waking up in the morning or after naps. Contrary to narcolepsy, patients have no REM-intrusion phenomena. Idiopathic hypersomnia is divided into

- *idiopathic hypersomnia with long sleep time*, where patients have EDS despite sleeping more than 10 hours per night, and
- *idiopathic hypersomnia without long time sleep*, where patients have EDS despite normal amounts of sleep, but less than 10 hours of sleep per night.

The pathophysiology of idiopathic hypersomnia remains elusive. Spinal fluid hypocretin levels are

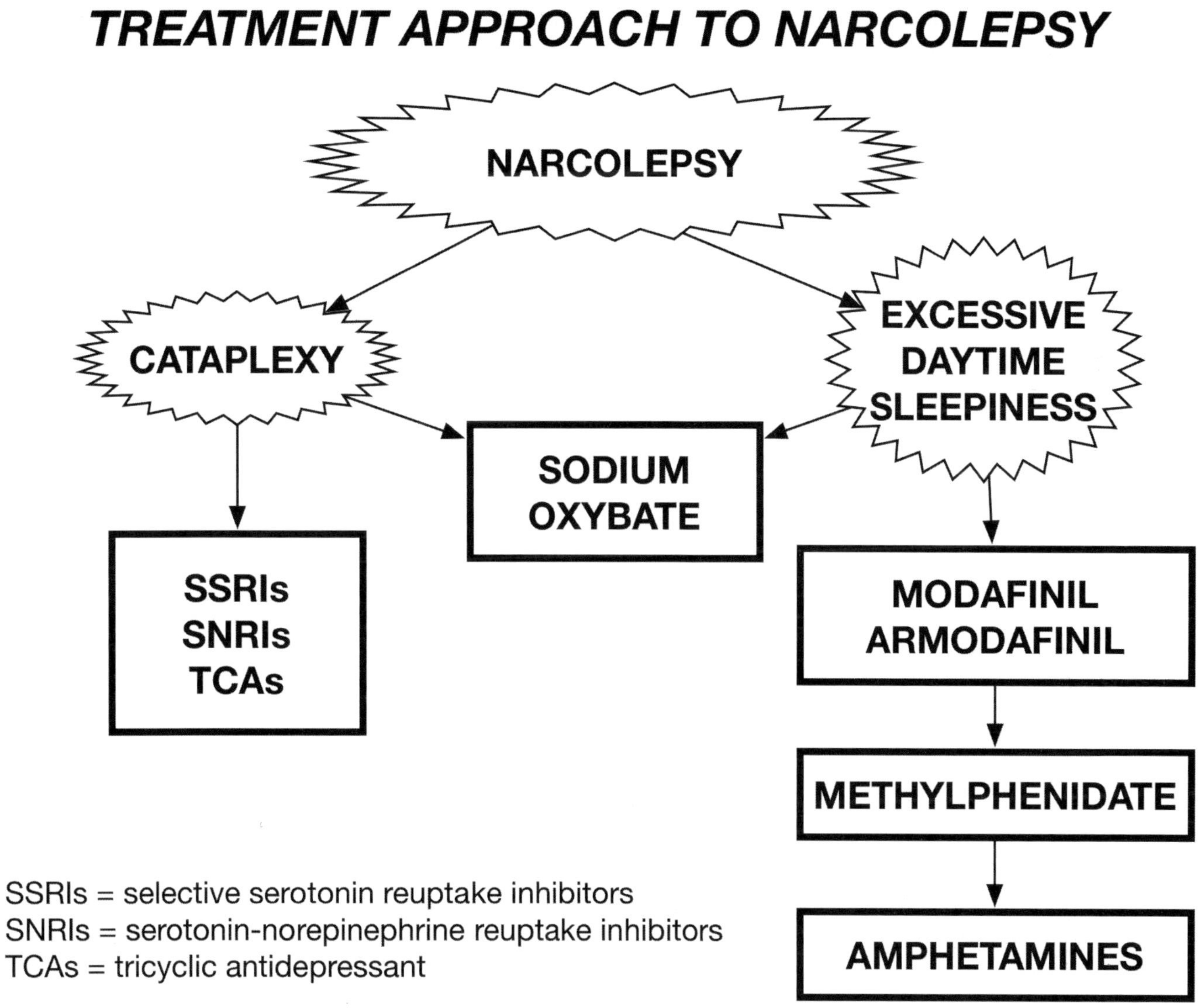

Figure 6-4. *Treatment approach to narcolepsy.*

FIGURE 6-5 NARCOLEPSY VS IDIOPATHIC HYPERSOMNIA		
	Narcolepsy	**Idiopathic Hypersomnia**
Naps are restorative	Yes	No
REM-intrusion phenomenon	Yes	No
Disrupted nocturnal sleep	Yes	No
Non-remitting	Yes	Sometimes remits

normal (>200 pg/mL). There is no association between idiopathic hypersomnia and human leukocyte antigen. Hypersomnia is an exclusion diagnosis after sleep breathing disorders, circadian rhythm disorders, insufficient sleep, disturbed nocturnal sleep disorder, narcolepsy, and abuse of sleeping-aid medications and alcohol are ruled out. **Figure 6-5** shows the main differences between narcolepsy and idiopathic hypersomnia. Management of EDS in idiopathic hypersomnia is the same as in narcolepsy.

Klein-Levin Syndrome, which affects teenage males, presents with recurrent episodes of excessive daytime sleepiness, hyperphagia, hypersexuality, and changes of cognition and mood. During the episodes, patients sleep more than twelve hours a day. Episodes last day to weeks and alternate with periods of normal alertness, cognition, and mood. Patients may respond to *lithium*, *valproic acid*, *lamotrigine*, or *amantadine*.

Menstrual-related hypersomnia is a rare syndrome, characterized by episodes of recurrent hypersomnia during menses, usually in the first few months after menarche. Oral contraceptives improve daily sleepiness, suggesting a hormonal imbalance. There is no specific therapy.

7

Circadian Rhythm Disorders

Circadian rhythm disorders are a complex set of sleep problems that involve the body's intrinsic timer. The word *circadian* originates from Latin roots – *circa*, meaning "about" and *dies*, meaning "day." The body's "internal clock," which in humans is approximately 24 hours, helps regulate the sleep/wake cycle.

Circadian rhythms are influenced by outside cues known as *zeitgebers*. Zeitgebers can shift the natural flow of our circadian rhythms. As we go through our day, we are "entrained" by these cues. Light is an example of a powerful zeitgeber. For example, when bright sunlight starts shining through our windows, we feel that it is time to get up out of bed to start the day. We also feel that when it gets dark out, the time for rest is near.

Circadian rhythm disorders can last a few days (e.g. jet lag) or be chronic. They occur when we are forced to break from our intrinsic routines to adjust to the normal time conventions set by society. Your body has its own built-in routine, and trying to change it can disrupt your sleep.

Pathophysiology of Circadian Rhythm Disorders

The main control center for the body's biological clock is located in the *suprachiasmatic nucleus* in the anterior hypothalamus. Specialized eye photoreceptors called melanopsin-containing retinal ganglion cells sense light and send signals via the *retinohypothalamic tract* to the suprachiasmatic nucleus. The suprachiasmatic nucleus then signals the pineal gland to increase or decrease secretion of the sleep-inducing hormone, *melatonin*. Under the control of the suprachiasmatic nuclei, the main enzyme that regulates the synthesis of melatonin is activated at night via sympathetic innervation of the superior cervical ganglia (**Fig. 7-1**). Light inhibits this pathway and decreases the production of melatonin. Darkness promotes melatonin production and sleep. *Lesions in the suprachiasmatic nucleus have been associated with disruption of circadian sleep patterns.*

There are several types of circadian rhythm sleep disorders:

- Delayed sleep phase syndrome
- Advanced sleep phase syndrome
- Jet lag disorder
- Irregular sleep/wake pattern
- Free-running pattern
- Shift work sleep disorder

Delayed Sleep Phase Syndrome

When thinking of a *delayed sleep phase*, it helps me to remember my brother. Although he sleeps around 8 hours a night, he would much rather go to bed at 2 am and wake up at 10 am. If he tries to sleep earlier than 2 am, he ends up just staring at the ceiling until that "magic" hour rolls around. By the same token, trying to wake him up at 6 am is nearly impossible. Alarm clocks, a gentle nudge, or even dousing him with a cup of water

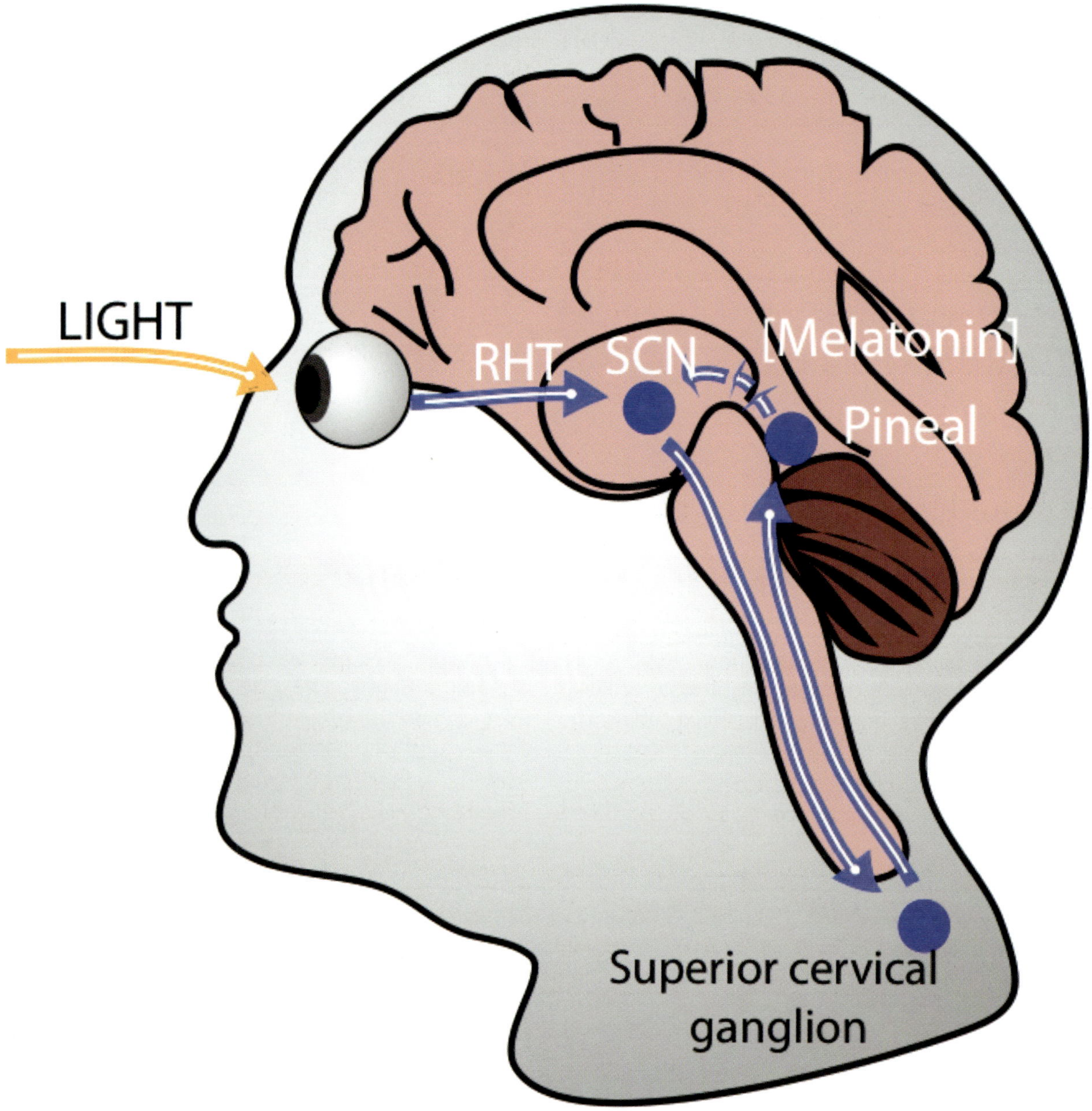

Figure 7-1. *Retinohypothalamic tract (RHT), suprachiasmatic nuclei (SCN), and pineal gland connections.*

does not seem to do the trick. As long as he keeps to his schedule, he does fine, but getting him to adopt a more "normal" schedule is difficult.

Advanced Sleep Phase Syndrome

"Early to bed, early to rise makes a man healthy, wealthy and wise." This saying by Benjamin Franklin is the perfect example of *advanced sleep phase syndrome*. People are in bed around 6-8 pm and wake up as early as 5 am or earlier. Like the delayed sleep phase syndrome, as long as these individuals are able to keep this schedule, they have no sleep problems. However, when people try to change to conform to the times set by society, they may have trouble with sleepiness and insomnia.

Jet Lag Sleep Disorder

Jet lag occurs when people travel across multiple time zones. For example, traveling from New York to California, there can be as much as a four hours time difference. A person in this scenario would have no trouble being wide awake at 8 am California time, because in New York, it's almost noon. However, when 8 pm in California rolls around, our native New Yorker is past his normal 10 pm bedtime and has trouble staying awake.

Irregular Sleep/Wake Pattern

Irregular sleep/wake pattern is a random sleep pattern where there is almost no consolidated sleep. Sleep is broken into a number of brief intervals that resemble those of experimental animals who have had damage to their suprachiasmatic nuclei. Sleep patterns like this can be found in patients with advanced Alzheimer's disease.

Free-Running Sleep Pattern

Free-running sleep pattern occurs because the length of the human circadian rhythm is actually a little longer than 24 hours. For sighted individuals in normal day-to-day conditions, the circadian rhythm is entrained to 24 hours by external cues such as daylight. The normal individual consistently goes to bed at the same time each day and arises at the same time each day. Free-running sleep pattern can be seen in blind individuals or in controlled areas such as in research environments where external cues are removed or in places under constant darkness, like a submarine. Individuals with this disorder go to sleep a little later each day and wake up a little later the following day, with the sleep/wake cycle progressively getting out of sync.

Shift Work Sleep Disorder

Shift work sleep disorder is a circadian rhythm disorder resulting from non-routine work schedules. When you think of this disorder, think of your doctors who sometimes have to go to the hospital in the middle of the night; your police officers and security guards who sometimes work the overnight shift one week, then work 9 to 5 the following week; or your friendly convenience store clerk who works from 11 pm to 7 am.

Figure 7-2 depicts the different circadian rhythm disorders.

Diagnosis of Circadian Rhythm Disorders

Circadian rhythm disorders can be diagnosed after other sleep disorders, medical conditions, medications, or toxins are excluded. Sleep diaries and actigraphy (**Chapter 2**) are helpful tools in the diagnosis. They provide a general overview of the sleep pattern. Measuring *core body temperature nadir* (*CT min*) and

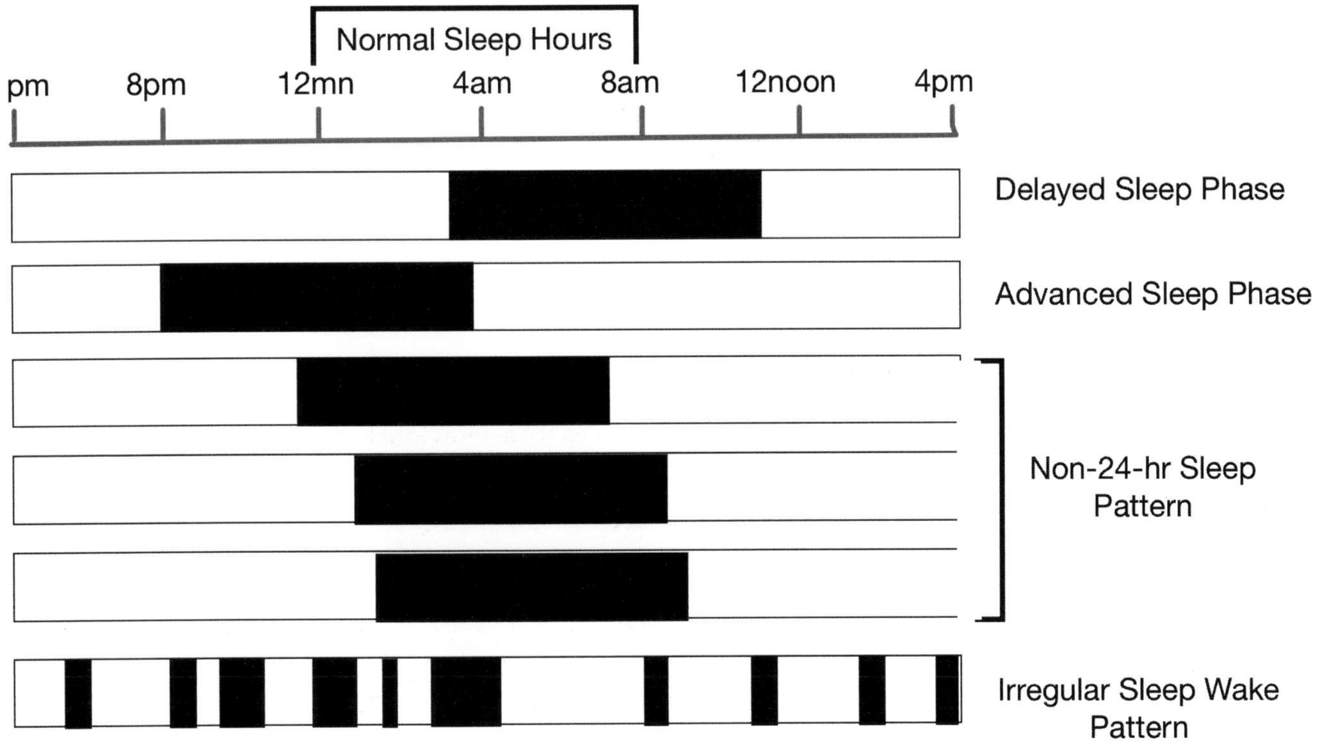

Figure 7-2. Graphic representation of different circadian sleep disorders.

melatonin samples in dim light (the *Dim Light Melatonin Onset test*) have been used only in research. CT min occurs 2-4 hours before awakening from sleep.

The dim light melatonin onset test of melatonin in saliva or plasma (or its metabolite 6-sulphatoxymelatonin in urine) under dim light conditions is the single most accurate marker of the circadian pacemaker. It represents the time when melatonin rises more than 3 and 10pg/ml for saliva and plasma, respectively. It takes place about 2-3 hours before bedtime. Dim light melatonin onset is mainly utilized to determine if an individual's circadian pattern is entrained to a 24-hour light/dark cycle or whether there is delayed or advanced circadian rhythm pattern, or a free-running state. Dim light melatonin onset is also used as a marker to determine the most appropriate time to use bright light therapy or exogenous melatonin (see below).

Treatment of Circadian Rhythm Disorders

Timed light exposure and *timed administration of melatonin* are the standard therapies for all circadian rhythm disorders. If a patient suffers from *advanced sleep phase* disorder, bright light may be used at night, and exogenous melatonin when the patient wakes up early at night, and wearing dark glasses following dawn. The converse applies with *delayed sleep phase* when exogenous melatonin is given before the desired bedtime and bright light is used in the morning to assist in the wakening process (**Fig. 7-3**). The bright light therapy consists of 5000 lux for 2 hours at about one meter from eye level. Sleep-inducing drugs and stimulants should be considered for shift work and jet lag disorders. *Physical exercise* may also be beneficial to increase body temperature, modify immune activity, reduce anxiety, and increase the serotonin level. Some evening exercise seems to help people to fall asleep faster and spend more time in deep sleep. However, high-intensity exercise less than one hour before bedtime delays falling asleep, with poorer sleep quality.

Night shift is a device used to treat night shift work sleep disorders. It consists of a device worn on the back of the neck that starts to vibrate when it senses that patients are falling sleep on their back and increases the intensity until position changes.

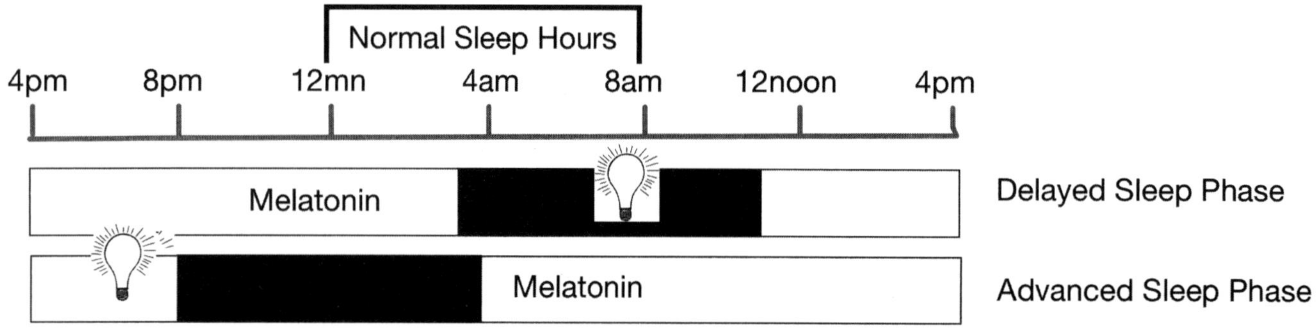

Figure 7-3. *Treatment of delayed sleep phase type and advanced sleep phase type. The light is used early in the morning (6-8 am) in delayed sleep phase type and in the evening (7-9 pm) in advanced sleep phase type. Melatonin is used 8-9 pm in delayed sleep phase type.*

8

Parasomnias and Complex Nocturnal Behaviors

Parasomnias are undesired behavioral or experiential phenomena that occur while falling asleep, during sleep, or while arising from sleep. The term *parasomnia* originates from the Greek prefix *para* meaning "alongside of" and from the Latin noun *somnus* meaning sleep. Parasomnias include a large number of medical disorders and are divided, according to the sleep state of origin, into *Non-REM Parasomnias, REM Parasomnias,* and *Other Parasomnias* (**Fig. 8-1**).

Non-REM Parasomnias

Non-REM parasomnias are behavioral disorders that result from a faulty arousal system. Patients experience incomplete arousal from sleep and display varying motor activities (sitting up, vocalization, walking) and emotions (fear, anger, sexual excitement), all without full awareness. Most Non-REM parasomnias in children are *confusional arousals, sleepwalking,* or *night terrors*. They share some features in common (**Fig. 8-2**).

Confusional arousals (sleep drunkenness) are recurrent episodes of nocturnal sleep arousal where the patient is confused and disoriented. Individuals may sit up, whimpering, crying, moaning, and saying words like "no" or "go away." Patients do not ambulate or have an activation of the autonomic nervous system with tachycardia, tachypnea, diaphoresis, flushing or mydriasis, as seen in sleepwalking and sleep terrors.

Sleepwalking (somnambulism) refers to recurrent behavioral episodes ranging from simple ones, where patients arise from sleep and wander around in the room or house, to more complex episodes such as cooking or cleaning.

Sleep terrors (pavor nocturnus) are recurrent Non-REMembered (unlike nightmares, which are remembered) dramatic episodes of sudden arousals with extreme panic, prominent motor activity (sitting up, looking afraid, inconsolable crying, vocalization or swearing, hitting the wall, running around the bedroom or house) and prominent autonomic nervous

FIGURE 8-1 TYPES OF PARASOMNIAS
Non-REM PARASOMNIAS
Confusional Arousals
Sleepwalking
Night Terrors
Sleep-Related Eating Disorder
Sexsomnia
REM PARASOMNIAS
REM-Behavior Disorder
Nightmare Disorders
OTHER PARASOMNIAS
Sleep-Related Dissociative Disorder
Sleep Enuresis
Sleep-Related Groaning
Exploding Head Syndrome

FIGURE 8-2 GENERAL FEATURES OF NON-REM PARASOMNIAS

- They are most common in children (first decade of life); occasionally adults.
- They dissipate during adolescence.
- They arise from Non-REM slow wave sleep (NREM3) – the first third of the night.
- There is a genetic predisposition.
- Events vary from one another in the same patient. They are not stereotyped as in epilepsy.
- They last 10 to 30 minutes.
- Patients are difficult to arouse during an event.
- During the events, patients are not fully conscious and do not respond to commands.
- Patients are amnestic or partially amnestic for the events.
- Events occur 3-4 times/week to every 1-2 months.
- The EEG shows slow-wave sleep activity and rhythmic theta or delta waves.

They are triggered by *exogenous factors* (drugs, alcohol, medications, environmental), *endogenous factors* (fever, menses, sleep deprivation), and sleep disorders of arousal, like obstructive sleep apnea, restless leg syndrome, and periodic leg movement disorder.

system activation (tachycardia, tachypnea, diaphoresis, flushing, and mydriasis).

Two forms of Non-REM parasomnias that present during adulthood are *sleep-related eating disorder* and *sexsomnia.*

Sleep-related eating disorder is characterized by recurrent episodes of involuntary eating or drinking during the night while asleep. Patients (usually young women) often consume high caloric foods or inappropriate foods (frozen pizza, cat food, toxic substances) with partial recollection of these events. This disorder results in sleep disruption, injuries, and weight gain. The disorder is associated with the use of the sedative *zolpidem (Ambien).*

Sexsomnia is characterized by sexual behaviors including fondling the partner, intercourse or attempts at intercourse, masturbation, sexual vocalization, and orgasm. It is more common in young adult males. It may be comorbid to sleep apnea, narcolepsy, and Parkinson's disease, and triggered by antidepressants, antiparkinsonian drugs, and shift work.

REM Parasomnias

The most common REM parasomnias are *REM-behavior disorder, sleep paralysis,* and *nightmare disorders.* These three syndromes share some features in common, as listed in **Figure 8-3.**

REM-behavior disorder consists of recurrent complex nocturnal behaviors with an aggressive motor activity (hitting, kicking), jumping from the bed, and vocalization (screaming) as part of a dream enactment. These nocturnal behaviors result from absence of the typical atonia (loss of tone) seen during normal REM sleep. This enables patients to act out their dreams in a typically agitated and violent manner, often resulting in injuries to patients or bedmates. REM-behavior disorder is a chronic disorder, commonly affecting men over age 50, who, during the attack, display behavior out of character for the patient during day hours. REM-behavior disorder is associated with the development of Parkinson's disease and Lewy body dementia and can be triggered by selective serotonin reuptake inhibitors or tricyclic antidepressants. REM-behavior disorder manifests during nocturnal polysomnography as increased chin EMG activity during REM sleep.

FIGURE 8-3 GENERAL FEATURES OF REM PARASOMNIAS

- Patients arise from REM sleep in the last half of the night or early morning hours soon before morning awakening.
- Patients usually arise fully alert and able to recall their experience.
- Events last minutes.
- Patients are unresponsive to the environment during the event.
- The arousal is abnormal for REM sleep.

Sleep paralysis occurs when normal REM-atonia presents during sleep onset or nocturnal awakening. Patients are fully conscious, but unable to move. Sleep paralysis may occur in isolation or as part of the narcolepsy complex (**Chapter 6**).

Nightmare disorder involves recurrent disturbing dreams that result in arousal from sleep. Individuals arise afraid, anxious, and fully alert, with complete dream recollection. Patients do not display any complex motor behaviors as seen in REM-behavior disorder.

Other Parasomnias

Sleep-related dissociative disorder is a dissociation episode (psychogenic spell) that occurs at night while the patient is actually awake by EEG, but is believed to be asleep. It resembles similar daytime episodes and presents with violence, sexual behavior, self-mutilation, screaming, walking, thrashing, screaming, bicycling movements, and running for minutes to hours. Sleep-related dissociative disorder is not strictly a parasomnia since the patient is awake, but rather a form of *pseudo-parasomnia.* It consists of the reenactment of a prior traumatic and violent experience during a traumatic or very stressful childhood.

Sleep enuresis is a recurrent involuntary voiding during sleep at least twice a week in a child older than

Figure 8-4. *REM-behavior disorder.*

five years of age. Enuresis can be divided into *primary enuresis*, where the child has never been consistently dry for six consecutive months before (genetic cause), and *secondary enuresis*, where the child has been dry for more than six consecutive months before. Secondary enuresis commonly is associated with sleep disorders of arousal, especially obstructive sleep apnea, diabetes, urinary infections, constipation, and stress (parents' divorce, death, disruptive families, neglect).

Sleep-related groaning (catathrenia) is an expiratory groaning during REM sleep and rarely during NREM. Sleep otherwise is normal and patients are asymptomatic, but not their roommates.

Exploding head syndrome is a sensation in the head of explosion or loud noise while falling asleep or upon awakening. It is more common in women. The syndrome may be triggered by stress or fatigue and typically remits over time.

Treatment of Parasomnias

The treatment of parasomnias is summarized in **Figure 8-5**.

Complex Nocturnal Behavior

Seizures in *nocturnal epilepsy* commonly occur during sleep, especially during Non-REM sleep, when brain electrical activity is more synchronized, and anti-gravitational muscle tone is still preserved enough to allow the expression of motor seizure activity. Nocturnal seizures can be confused with different parasomnias or sleep disorders that significantly disrupt the sleep structure and may have detrimental daytime consequences.

Nocturnal frontal lobe epilepsy is an epileptic disorder with nocturnal seizures that closely resemble disorders of arousal. Seizures are recurrent and stereotyped, with the abrupt onset of agitated, disorganized, or violent body movements involving the trunk and proximal limb muscles, body rocking, kicking, boxing, thrashing, pedaling, bending, hitting, spitting, running, and vocalizations. Episodes are brief ($<$ minute), and awareness is commonly preserved, which also differentiates them from parasomnias. Some cases have a hereditary background. The EEG may show seizure activity, although not always if the epileptic focus is not accessible to superficial EEG electrodes, or muscle

FIGURE 8-5 TREATMENT OF PARASOMNIAS	
GENERAL GUIDELINES	• Educate and reassure patients and parents. • Encourage sleep hygiene techniques. (**Fig. 5-6**) • Keep regular sleep hours. • Modify or eliminate triggers such as sleep breathing disorders, periodic leg movement disorder, and sleep deprivation. • Improve the safety of the sleep environment. • Consider cognitive behavioral therapy.
NON-REM PARASOMNIAS	• The majority do not require treatment. Episodes are self-limited with low injury risk and limited daytime recollection of the episodes. • Awaken patients 15-20 min before the expected episode (estimated time when they most commonly happen) to reduce the episode appearance. • Pharmacologic treatment is indicated when symptoms are present for more than six months and there is an associated risk of injuries. • Short half-life sedatives are indicated for events occurring in the first third of the night, such as - Triazolam 0.125-0.5 mg po qhs - Zolpidem 5-10 mg po qhs • Longer half-life sedatives are indicated for events occurring in the first half or throughout the night: - Lorazepam 1-2 mg po qhs - Clonazepam 0.5-2 mg po qhs Longer half-life sedatives carry more risk for daytime side effects. • Treat comorbid sleep disorders, i.e. obstructive sleep apnea, restless leg syndrome, and periodic leg movement disorder.
REM PARASOMNIAS	• Pharmacologic treatment is indicated when symptoms are present for more than six months and are associated with a risk of injuries. • Clonazepam 0.5-2 mg po qhs. However, clonazepam or other benzodiazepines are not recommended if the patient is demented, ataxic, or suffers from obstructive sleep apnea. • Alternatives to benzodiazepines include pramipexole 0.5-1 mg po qhs and melatonin 3-15 mg po qhs. • REM suppressant drugs such as tricyclics and clonidine could be considered if benzodiazepines fail or are contraindicated, as in obstructive sleep apnea.

FIGURE 8-6 AROUSAL DISORDERS VS NOCTURNAL FRONTAL LOBE EPILEPSY

	Arousal Disorders	Nocturnal Frontal Lobe Epilepsy
Age of onset	First decade	First or second decade
Times of occurrence	NREM3	NREM1 and/or NREM2
Duration	$>$ 1 min	$<$ 1 min
Stereotyped (similar) episodes	No	Yes
Cluster presentation (one after another)	Rare	Most likely
Daytime	No events	Events may occur
Onset/Offset	Gradual (over a few minutes)	Sudden
Awareness	Impaired	Commonly preserved
Postictal confusion	Present	Absent
EEG findings	Slow waves	Normal or epileptic activity
Treatment	• Cognitive behavioral therapy • Benzodiazepines	Anti-epileptics

activity during the seizures hides the visualization of epileptic discharges. Treatment consists of anti-epileptic medications such as *carbamazepine*. If the episodes occur only at night, anti-seizure medication could be taken before going to sleep. **Figure 8-6** shows the main differences between arousal disorders and nocturnal frontal lobe epilepsy.

9

Restless Leg Syndrome and Periodic Limb Movement Disorders

Restless Leg Syndrome

Restless leg syndrome is a sensorimotor neurological disorder characterized by four key features:

- There is an urge to move the legs, usually associated with an uncomfortable and unpleasant feeling in the legs.
- It improves with activity.
- It worsens with inactivity.
- It is worse at night.

These four features are all required to make the diagnosis.

Restless leg syndrome is common; it affects 10% of adults in the United States, mainly Caucasians, compared to African Americans and Asians. In adults, women are more affected than men, but in children the incidence is equal in both sexes. It is chronic, although remission is reported rarely, especially in its secondary forms (see below). Restless leg syndrome usually affects both legs, but it may affect just one leg and/or the upper limbs. 80% of the patients with restless leg syndrome also have *periodic leg movement during sleep*.

There are two forms of restless leg syndrome: *primary* and *secondary*.

Primary restless leg syndrome occurs independently of other disorders, although it may be aggravated or triggered by some factors. It has a strong genetic background, and more than half of the patients have a positive family history. Several genes have been implicated with the primary form, most of them with an autosomal dominant inheritance.

Secondary restless leg syndrome is precipitated by other disorders that, when properly treated, result in resolution or improvement of the syndrome. **Figure 9-1** lists secondary causes of restless leg syndrome.

Diagnosis of Restless Leg Syndrome

Figure 9-2 displays the main features in the diagnosis of restless leg syndrome in adults and adolescents. **Figure 9-3** shows supportive evidence for restless

FIGURE 9-1 SECONDARY CAUSES OF RESTLESS LEG SYNDROME
• Iron deficiency anemia • End-stage renal disease • Pregnancy • Medications (antidepressants, antipsychotics, dopamine antagonists, antihistamines) • Peripheral neuropathies • Radiculopathies • Fibromyalgia • Multiple sclerosis • Parkinson's disease • Folate and magnesium deficiency

FIGURE 9-2 MAIN FEATURES OF RESTLESS LEG SYNDROME IN PATIENTS > 12 YEARS OLD
• The urge to move the legs may or may not be associated with uncomfortable sensations in the legs. • Symptoms begin or worsen during periods of inactivity, whether lying or sitting. • Symptoms are partially or completely relieved by movement — moving or stretching the legs or walking. • Symptoms worsen in the evening and at night.

FIGURE 9-3 SUPPORTIVE EVIDENCE FOR RESTLESS LEG SYNDROME
• Family history (first degree relatives with restless leg syndrome) • Positive response to dopaminergic therapy • Periodic leg movements during sleep

FIGURE 9-4 DIAGNOSIS OF RESTLESS LEG SYNDROME IN CHILDREN	
A alone, or **B** + **C** satisfy the criteria	
A	The child in his own words meets the adult criteria for RLS.
B	The child meets adult criteria for RLS, but not in his or her own words.
C	The child has at least two of the three following findings: • A sleep disturbance for age • A biological parent or sibling with definite restless leg syndrome • A nocturnal polysomnography documenting a periodic leg movement index of 5 or more movements per hour of sleep

leg syndrome, which, although not necessary for the diagnosis, contributes to it.

In children, restless leg syndrome is more difficult to diagnose, since children have more difficulty communicating what they feel (**Fig. 9-4**).

Restless leg syndrome may be confused with other nocturnal disorders that have significant sensory or motor activity, but specific features point to the right diagnosis (**Fig. 9-6**).

Pathophysiology of Restless Leg Syndrome

Restless leg syndrome is probably caused by dysfunction of the dopaminergic system (decreased dopamine receptors, changes in dopamine receptor binding and dopamine uptake). Deficiency of iron in the brain is also suspected. Iron is required in the rate-limiting enzyme in dopamine production.

Figure 9-5. Restless leg syndrome.

FIGURE 9-6 DIFFERENTIAL DIAGNOSIS OF RESTLESS LEG SYNDROME (RLS)		
DISORDER	**LIKE RLS**	**UNLIKE RLS**
Neuroleptic (antipsychotic drug)-induced akathisia	Inability to remain still with a general restlessness	• Movement affects several body parts with no circadian pattern. • Low incidence of periodic leg movements during sleep • Associated with the use of neuroleptics
Anxiety	Leg movements and increased motor activity	• No sensory discomfort • Increased sympathetic nervous system activity with sweating and palpitation • Responds to benzodiazepines
Positional leg discomfort	Leg discomfort after being in one position for awhile, relieved with movement or after changing body positions	• No urge to move the legs • No circadian rhythm • Pain may be relieved with movement, but usually after more complicated exercises than simple leg movements.
Nocturnal cramps	Nocturnal leg pain (calf), occurs at night during rest, and is relieved with activity.	• No restlessness • Focal muscle hardening (palpable tightening), usually affecting calf muscles • Relieved after stretching rather than just moving the leg(s) • No response to dopamine agonists
Peripheral neuropathy and/or radiculopathy	Leg discomfort, worse at night when patient lies down in bed	• No restlessness • No improvement with movement • Presence of other neurologic features • No response to dopamine agents
Painful leg and moving toes syndrome	Discomfort in legs (one or both) associated with constant and involuntary writhing movements of the toes	• Movements do not improve sensory complaints. • Constant involuntary writhing movement of the toes • Repetitive • No circadian pattern
Rhythmic movement disorder in children	Head banging, head rolling, body rolling, and body rocking. When it mainly affects legs, it may be confused with RLS.	• Repetitive, stereotypical, and rhythmic movements that start before going to sleep and continue into light sleep

Treatment of Restless Leg Syndrome

Treatment for restless leg syndrome relies on three basic principles:

1. Avoid triggers such as caffeine, alcohol, and tobacco.
2. Treat secondary causes, such as iron deficiency; eliminate medications that trigger or aggravate restless leg syndrome.
3. Use drugs that treat restless leg syndrome, such as dopamine agonists, L-dopa/carbidopa, pregabalin, gabapentin, clonazepam, and opioids (**Fig. 9-7**).

The first-line treatment for restless leg syndrome without neuropathy is a *dopamine agonist* such as *ropirinole, pramipexole,* or *rotigotine.* If patients do not respond to the first dopamine agonist, another dopamine agonist should be considered before changing to another drug family. Pramipexole titration is slower (every two weeks) in patients with renal disease. *L-dopa-carbidopa* is no longer recommended as first-line therapy; its use should be limited to no more than two nights per week to prevent a common complication called *augmentation.* Augmentation consists of worsening of restless leg syndrome, which becomes more severe, arising sooner in the day or affecting other parts of the body not originally affected, such as the arms. Augmentation is seen in about 80% percent of patients on L-dopa/carbidopa and a third or less of patients taking dopamine agonists. Treatment of augmentation consists of reducing the dose of the offending medication if mild, or changing to another drug family if more severe.

In patients where restless leg syndrome coexists with painful peripheral neuropathy or radiculopathy, *pregabalin, gabapentin enacarbil,* or *gabapentin* may be

FIGURE 9-7 TREATMENT OF RESTLESS LEG SYNDROME		
GENERAL MEASURES	• Avoid caffeine • Mild exercises • Avoid alcohol • Circadian adjustments • Hot or cold bath (may help) • Discontinue tobacco	
TREAT SECONDARY CAUSES	• Discontinue neuroleptics: haloperidol, phenothiazine derivatives, if possible. • Discontinue tricyclics and serotonin antidepressants, if possible; bupropion is safe because it has dopaminergic activity. • Discontinue antiemetics: prochlorperazine or metoclopramide. • Discontinue cold and allergy medications. • Ferrous sulfate 325mg with vitamin C 100mg BID or TID if ferritin levels are <50 ug/L or ng/ml. Low ferritin may trigger or aggravate RLS.	
Medication	**Dosage**	**Adverse effects**
Ropirinole	0.25mg 2 to 3 hours before bedtime; dose may be increased by 0.25mg every 3-5 days up to 2-4mg.	Nausea, vomiting, fluid retention, nasal stuffiness, orthostatic hypotension, lightheadedness and daily drowsiness
Pramipexole	0.125mg 2 to 3 hours before bedtime; dose may be increased by 0.125mg every 3-5 days up to 1mg.	Nausea, vomiting fluid retention, nasal stuffiness, orthostatic hypotension, lightheadedness and daily drowsiness
Rotigotine	1mg patch once a day; dose may be increased by 1mg every week up to 3 mg every day.	Nausea, skin reactions, headaches
L-dopa/carbidopa	25/100mg at bedtime; dose may be increased as needed. Recommended use no more than 2 nights per week. Nightly use may cause augmentation (worsening) in most patients.	Nausea, vomiting, fluid retention
Pregabalin	75mg at bedtime, may be increased to 150mg, max dose 450mg divided into 2 nightly doses.	Drowsiness, dizziness, peripheral edema
Gabapentin	300mg at bedtime to be increased 300mg weekly to efficacy. Dose may be increased up to 2400mg divided in 2 to 3 nightly doses.	Drowsiness, dizziness, peripheral edema
Gabapentin enacarbil	600mg at bedtime	Drowsiness, dizziness, peripheral edema
Clonazepam	0.5mg at bedtime to be increased gradually up to 2mg	Excessive sedation, dependency
Methadone	5 mg at bedtime; dose may be increased weekly up to 40 mg/d.	Nausea, vomiting, constipation, hypotension, dizziness, asthenia, sedation
Oxycodone	5 mg at bedtime; dose may be increased by 5 mg per week up to 20mg/d	Excessive sedation, constipation; significant abuse potential may limit usefulness.

considered as first-line therapy. Otherwise, pregabalin, gabapentin, and opioids are usually considered second-line medications. The doses of pregabalin, gabapentin enacarbil, and gabapentin need to be adjusted to renal function since they are released via the kidneys.

Several opioids have been used in the therapy of restless leg syndrome, including *methadone, oxycodone, morphine, codeine* and *tramadol*. The development of dependence is low, but is still a concern. They worsen obstructive sleep apnea, if coexistent.

Clonazepam has no effect on reducing restless leg movements or periodic leg movements during sleep (see below). Its main efficacy is as a sedative that provides faster sleep onset and fewer periodic leg movements during sleep-arousals. Clonazepam is associated with cognitive problems and increased risk for falls.

Different devices are under investigation as non-drug therapy for restless leg syndrome. The first group of devices increases blood circulation via whole body vibration, pneumatic compression, or near-infrared light. Vibratory counter-stimulation via transcutaneous electricity or vibration are other types of devices. Only vibratory counter stimulation (*Relaxis*) is FDA approved to treat restless leg syndrome.

Similar therapy may be used in children with restless leg syndrome.

Figure 9-8 summarizes the treatment approach to restless leg syndrome.

Periodic Limb Movement During Sleep

Periodic limb movement during sleep is characterized by periodic (repetitive and rhythmic), stereotyped leg movements with big toe extension, ankle flexion (dorsiflexion), and knee flexion. It is involuntary, occurring while asleep, as opposed to restless leg

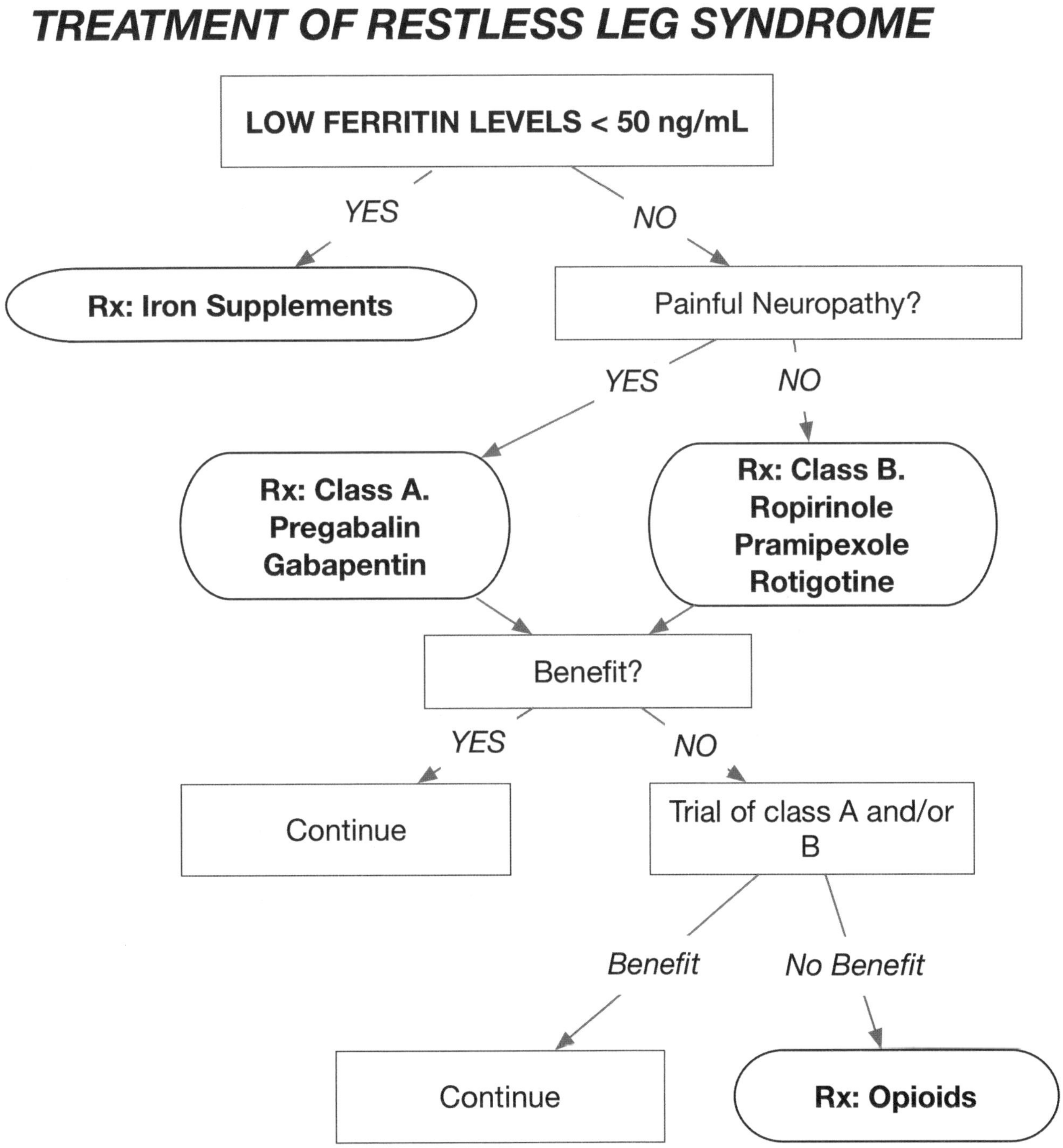

Figure 9-8. Treatment of restless leg syndrome.

syndrome, which keeps the patient awake! This disorder rarely may affect the upper limbs, with flexion of the elbow. When periodic limb movement during sleep is accompanied by sleep disturbances, not attributed to another sleep disorder, it is called *periodic limb movement disorder*. If sleep gets significantly fragmented, patients may complain of non-refreshing sleep, excessive daytime sleepiness, and fatigue. Periodic leg movement may occur during sitting or lying during restful wakefulness, which is called *periodic leg movement during wakefulness*.

Periodic limb movement during sleep affects 5% of the population, males and females equally, and is more common in middle-aged and older adults. Periodic limb movement during sleep shares the same risk factors as restless leg syndrome. 80% of the patients with restless leg syndrome also suffer from periodic limb movement during sleep. However, the reverse is not true, and most of the patients with periodic limb movement during sleep do not have restless leg syndrome. Periodic limb movement during sleep coexists with other sleep disorders like narcolepsy, REM-behavior disorder, and obstructive sleep apnea.

Diagnosis of Periodic Limb Movement During Sleep

Periodic limb movement during sleep requires polysomnography to confirm the diagnosis, while restless leg syndrome requires only a history. **Figure 9-9** displays differences between restless leg syndrome and periodic leg movement disorder. **Figure 9-10** shows the main polysomnography features of period leg movement during sleep. Polysomnography uses electromyography leads attached to the anterior tibialis muscle. *The Periodic Limb Movement Index (PLMI)* represents the number of periodic limb movements per hour of total sleep time. The PLMI is abnormal when it is > 15 in adults and > 5 in children. Most periodic limb movements during sleep occur in the first part of the sleep.

Treatment of Periodic Limb Movement During Sleep

Therapy for periodic limb movement during sleep is similar to the therapy for restless leg syndrome.

FIGURE 9-9 RESTLESS LEG SYNDROME VS PERIODIC LEG MOVEMENT DURING SLEEP

	Restless leg syndrome	**Periodic leg movement during sleep**
Status	Awake	Asleep
Diagnosis	History	Polysomnography
Presenting sleep complaint	Sleep-onset insomnia	Sleep-maintenance insomnia or excessive daytime sleepiness
Patient awareness	Yes	No

FIGURE 9-10 POLYSOMNOGRAPHY IN THE DIAGNOSIS OF PERIODIC LEG MOVEMENT DURING SLEEP

- Duration of muscle movement is 0.5-5 seconds.
- Muscle movements occur in a series of 4 or more consecutive leg movements (isolated movements are not counted).
- Minimum time of 5 seconds between onset of consecutive muscle movements.
- Maximum time of 90 seconds between onset of consecutive muscle movements.

NOTE: Typical duration between consecutive muscle movements is 20-40 seconds.

- An EMG amplitude equal or > 25% from the baseline foot dorsiflexion noted during calibration.
- Movement does not occur within 0.5 seconds of a respiratory event (apnea or hypopnea).

10

Sleep Disorders In Children, Women, and The Elderly

Sleep Disorders in Children

The study of childhood sleep disorders requires an understanding of normal pediatric sleep. Sleep architecture and sleep requirement in hours change from birth to adolescence.

Normal *newborns* (first month of life) spend 16-20 hours per day sleeping, in periods of 1-4 hours, equally distributed across the day and night. The distinction between awake and asleep is mainly behavioral: newborn is *awake* if eyes are open, and *asleep* if eyes are closed for more than a minute. The differentiation among sleep stages becomes difficult in the absence of specific EEG waves (not developed yet). Sleep time is divided between *"active sleep"* (REM-sleep equivalent), *"quiet sleep"* (NREM-sleep equivalent), and *"intermediate sleep,"* a sleep that is neither active nor quiet. Active sleep is the predominant form of sleep-onset and represents half of the sleep time. Quiet sleep is characterized by the sleep architecture of *trace alternant* (burst of high amplitude mixed frequencies alternating with a period of electrical silence on EEG).

Figure 10-1 shows the main characteristics of active and quiet sleep.

During *infancy* (1 month to 1 year old):

- Total sleep time decreases to 12-16 hours, including naps. During this life period, children are able to sleep throughout the night, while still having two daily naps of 2 to 4 hours each.
- Before month 6, sleep is distributed equally across the day and night. Sleep is divided between active, quiet, and intermediate; however, the appearance of waves such as spindles by month two and K-complexes by month four allows the early recognition of NREM and REM sleep.

FIGURE 10-1 DIFFERENCES BETWEEN ACTIVE AND QUIET SLEEP

Main Features	Active	Quiet
Rapid eye movements	Present	Absent
Body movements	Abrupt	Absent or minimal
Breathing	Irregular	Regular
Trace alternant on EEG	Absent	Present

- By month 6, sleep starts to be more consolidated at night. By month 9, two thirds of infants sleep 6-8 hours per night, almost all by month 12. During the second half of infancy, all NREM stages 1,2,3 can be differentiated, and NREM-REM sleep cycles are repeated every 50 minutes, rather than 90-120 minutes, as in adults.

Figure 10-2 shows the main features of sleep during infancy.

Toddlers (age 1-3 years) and *preschoolers* (age 3-5 years) sleep 1 to 3 hours less than infants. Toddlers still nap (1-3 hours).

By *age 5 years*, REM sleep is reduced to 20-22% of total sleep time, and daily naps disappear.

School-age children (age 6-12 years) sleep 10-11 hours per day, and sleep routine accommodates to school hours. Daily sleepiness is rare in school-age children.

During *adolescence* (age 13-18 years) total sleep time is reduced to 9 hours, and sleep cycles are delayed, with the tendency to later sleep and waking hours.

Sleep disorders may affect one in four children and change according to age. During infancy, sleep is normally fragmented during the first year of life, so the distinction between normal and abnormal sleep remains unclear. After infancy, children share similar sleep disorders as adults, but with some specific peculiarities, according to age. The most common pediatric sleep disorders are

- insomnia
- circadian rhythm disorders
- parasomnias
- sleep breathing disorders

Regardless of the particular sleep disorder, the resulting sleep deprivation, caused by lack of continuity or duration of sleep, generates changes of mood, behavior, and cognition that result in poor social and school performances.

Keep in mind that insufficient sleep is the leading cause of daily drowsiness in children and adolescents rather than sleep disorders. The recommended hours by age of sleep to prevent health malfunction are shown in **Figure 10-3**. (The American Academy of Sleep Medicine did not make specific recommendations for newborns 0-3 months because of the wide range of variation in sleep duration and patterns.)

The approach to sleep disorders in children starts with the *history*. The most commonly used questionnaire is the *BEARS questionnaire*. This is an acronym that stands for **B**: Bedtime problems, **E:** Excessive daytime sleepiness, **A:** Awakening at night, **R:** Regularity and duration of sleep, **S:** Snoring (**Fig. 10-4**).

Physical examination focuses on growth parameters, dysmorphic body features (e.g. craniofacial abnormalities), and examination of the lungs, heart, oropharynx (tonsils and adenoids), and nervous system (strength, physical and cognitive milestones).

Insomnia in Children

Insomnia is the inability to initiate or maintain sleep, or early morning awakening despite an adequate

FIGURE 10-2 MAIN SLEEP FEATURES DURING INFANCY

Main Features	< 6 months	> 6 months
Sleep distribution	Equally across day and night	More at night
Sleep stages	Active sleep, quiet sleep, and intermediate sleep	NREM stages 1,2,3 and REM
Sleep onset stage	Active sleep (REM)	NREM
Predominant sleep stage	REM	NREM

FIGURE 10-3 SLEEP DURATION RECOMMENDATIONS BY AGE
from the American Academy of Sleep Medicine and National Sleep Foundation
*** including naps**

Age	Recommended hours	Minimal	Maximal
Newborn (first month)	14-17 *	11-13 *	18-20 *
Infants (1-12 months)	12-16 *	10-11 *	17-18 *
Toddler (1-3 years)	11-14 *	9-10 *	15-16 *
Preschoolers (3-5 years)	10-13 *	8-9 *	14 *
School-age (6-12 years)	9-12	7-8	12
Adolescents (13-18 years)	8-10	7	11

FIGURE 10-4 BEARS SLEEP SCREENING ALGORITHM (P) Parent-directed question; (C) Child-directed question, from *A Clinical Guide to Pediatric Sleep: Diagnosis and Management of Sleep Problems* by Jodi A. Mindell and Judith A. Owens; Lippincott Williams & Wilkins			
	Toddler/ Preschool (2-5 years)	**School-aged (6-12 years)**	**Adolescent (13-18 years)**
1. **B**edtime problems	Does your child have any problems going to bed? Falling asleep?	(P) Does your child have any problems at bedtime? (C) Do you have any problems going to bed?	(C) Do you have any problems falling asleep at bedtime?
2. **E**xcessive daytime sleepiness	Does your child seem overtired or sleepy during the day? Does she still take naps?	(P) Does your child have difficulty waking in the morning, seem sleepy during the day, or take naps? (C) Do you feel tired a lot?	(C) Do you fall sleep a lot during the day? In school? While driving?
3. **A**wakenings during the night	Does your child wake up a lot at night?	(P) Does your child seem to wake up a lot at night? Any sleepwalking or nightmares? (C) Do you wake up a lot at night? Have trouble getting back to sleep?	(C) Do you wake up a lot at night? Have trouble getting back to sleep?
4. **R**egularity and duration of sleep	Does your child have a regular bedtime and wake time? What are they?	(P) What time does your child go to bed and get up on school days? Weekends? Do you think he/she is getting enough sleep?	(C) What time do you usually go to bed on school nights? Weekends? How much sleep do you usually get?
5. **S**noring	Does your child snore a lot or have difficulty breathing at night?	(P) Does your child have loud or nightly snoring or any breathing difficulties at night?	(P) Does your teenager snore loudly or nightly?

opportunity to sleep, associated with daily dysfunction for at least three months.

In children, as opposed to adults, insomnia is often reported by caretakers as bedtime resistance or inability to sleep independently.

In *sleep-onset association insomnia,* a condition is required to fall sleep and to return to sleep. Without the condition, sleep is delayed and nocturnal awakenings increase. These conditions are the presence of a caregiver in the bedroom, being held or rocked, watching TV, or having a specific object such as a Teddy bear or a bottle. This form of insomnia is common in young children and is not considered a true disorder until it significantly disrupts the sleep of the child and parents. Risk factors for this type of insomnia are co-sleeping, anxiety, and medical disorders.

Limit-setting insomnia is characterized by the resistance or refusal of the child to go to sleep at the appropriate bedtime. This results from the caregiver's inconsistent enforcement of bedtime at a specific time and/or when a child wakes up in the middle of the night.

These behavioral insomnias affect 10-30% of infants and toddlers. Their treatment consists of establishing regular sleep hours; making the bedroom suitable to sleep; putting children into bed when they are drowsy but still awake, to learn to make the association between the bedroom and sleep; and the use of the behavioral technique called *extinction,* where a child is put to sleep and suddenly or gradually ignored (the caretaker is less responsive to the child's request for assistance to fall asleep) until the child learns to fall asleep.

Older children and adolescents can also suffer from insomnia. *Inadequate sleep hygiene insomnia* is the most

Figure 10-5. *Limit-setting insomnia.*

common insomnia in this age range. It results from more autonomy with inappropriate bedtime hours, use of electronic devices at bedtime, consumption of coffee during the afternoon or evening hours, drug or substance abuse, and psychiatric disorders, such as anxiety and/or depression. The treatment requires sleep hygiene, maintaining consistent sleep hours, avoiding the nighttime use of electronic devices, and eliminating high consumption of coffee. Sleep aid drugs may be required. See **Chapter 5** for more information about insomnia.

Circadian Rhythm Disorders in Children

Delayed circadian rhythm disorder is the most common circadian rhythm disorder of adolescence. It affects 7-16% of teenagers. In this disorder, the sleep/awake hours are delayed by two or more hours from the desired bedtime and waking time respectively. Patients present with sleep-onset insomnia and difficulty arising in the morning and conforming to society's normal sleep/wake hours. Patients are commonly drowsy during the daytime, but wake up refreshed, if sleep is uninterrupted. School avoidance, anxiety, and depression are common. Delayed circadian rhythm disorder runs in families. Diagnosis requires a detailed history and logs of the sleep/wake hours. Polysomnography is not indicated unless a sleep breathing disorder is suspected. Treatment is based on proper sleep hygiene: avoiding stimulant activities at night such as intense physical activity, use of electronic devices, and caffeine intake; eliminating daily naps. In refractory cases, *chronotherapy* is required. This technique consists of progressive daily change in the sleep and waking hours by 2-3 hours until the target sleep time is attained and sustained, usually within a week. This technique requires motivation since it needs to be maintained during weekends and vacation. Also consider melatonin at bedtime to induce sleepiness and light therapy in the morning to assist in awakening.

Parasomnias in Children

Parasomnias are abnormal and unpleasant behavioral, physical (motor, vocal), autonomic phenomena while falling asleep, during sleep, or in the transition from sleep into wakefulness. Parasomnias are not always associated with sleep disruption and daytime sleepiness. They are divided into *sleep-stage dependent* (Non-REM parasomnias and REM parasomnias) and *non-sleep-stage dependent*.

Non-REM parasomnias occur during Non-REM slow-wave sleep during the first third of the night. They primarily affect toddlers and schoolchildren, run in families, and are triggered by specific factors (**Fig. 10-6**). They present with abnormal arousals with restoration of motor activity, without full consciousness. There is altered judgment, and the child is difficult to awaken. If aroused, the child is unaware of what happened and rapidly returns to sleep, not remembering the event. There are three Non-REM parasomnia disorders, which clinically overlap in their features:

- sleepwalking
- confusional arousals
- sleep terrors

In *sleepwalking,* children arise from sleep with stereotypical wandering around, acting confused, strange, calm, and without signs of fear. These events range from simply sitting up in bed or walking around the house to complex behaviors, such as eating and talking.

In *confusional arousals,* children arise from sleep with confusion, disorientation, and are more agitated than in sleepwalking. If forced arousal is attempted, children became combative, resistant, and aggressive.

In *sleep terrors*, there is partial and dramatic arousal from sleep with behavioral and physical manifestation of intense fear, including panicky screaming, crying, and autonomic hyperactivity, with tachycardia, sweating, flushing, tachypnea, and muscle tension.

Treatment of Non-REM parasomnias focuses on reassurance, sleep environment safety, and elimination of trigger factors. Medications such as benzodiazepines, tricyclics, and specific serotonin reuptake inhibitors are rarely required.

REM parasomnias are a group of disorders characterized by awakenings from REM sleep with distress and inability to resume sleeping. Nightmares and sleep paralysis are examples of REM parasomnias.

FIGURE 10-6 NON-REM PARASOMNIA TRIGGER FACTORS
Sleep deprivation
Poor sleep hygiene
Sleep environment not suitable for sleep
Fever
Stress
Medications
- Sedative hypnotics
- Neuroleptics (antipsychotic drugs)
- Stimulants
- Antihistamines
Sleep disorders
- Obstructive sleep disorders
- Periodic Leg Movement of Sleep
Comorbid medical disorders
Comorbid neurological disorders
Comorbid psychiatric disorders

Nightmares are frightening dreams that awaken children from REM sleep with a detailed recollection of the frightening experience. Children wake up anxious, stressed, and afraid, with difficulty resuming sleep. Nightmares are more common in the second half of the night. They peak at age 6-10 years and may affect 10-50% of children. Risk factors for nightmares are traumatic events, anxiety, sleep deprivation, and drugs (antidepressants, antihypertensive medications, and dopamine agonists). Nightmares differ from sleep terrors, since nightmares awaken children from sleep with clear recollection of the dream and difficulty resuming sleep. Treatment focuses on reassurance, good sleep hygiene, and maintaining a low level of light at night. In severe cases, nightmares may require psychotherapy.

Sleep paralysis is a generalized weakness with preservation of consciousness, accompanied by distress, during the transition between sleep and wakefulness. Sleep paralysis lasts a minute or longer. Risk factors are stress, fatigue, sleeping supine, and sleep deprivation. It may occur in isolation (only feature) or as a component of the narcoleptic syndrome (see **Chapter 6**). Treat with reassurance.

Non-sleep state-dependent parasomnias occur during any sleep stage. The most common one is *enuresis*. Enuresis is bed wetting (urinary incontinence) while asleep, but not during daytime. It affects 25% of boys and 15% of girls at age 6. The criteria for diagnosis include two or more wet events per month at age 5-6 years, or one or more wet events per month after age 6 years, not explained by any other physical disorder (diabetes, bladder dysfunction, urinary infection, constipation, or seizures).

Enuresis is divided into *primary enuresis* (child was never continent) and *secondary enuresis* (child was continent for more than one year before becoming incontinent). Triggers of enuresis are social and psychological stresses. Treatment starts with good family support and no punitive behavior from the parents. There should be good hydration in the morning and early afternoon, with limited fluid intake at later hours. The combination of frequent and scheduled voids is recommended. If these measures fail, consider *alarm therapy* and *desmopressin*.

Alarm therapy consists of wet sensors attached to the underwear and activated by increased wetness. It alerts the child and parents to take the child to the bathroom to finish voiding. It is effective, but requires motivation from children and parents.

Desmopressin is an analog of vasopressin (antidiuretic hormone) that reduces the nocturnal urinary production. It may be used intermittently (camping, sleepovers) or chronically. Hyponatremia and seizures are reported as side effects and require monitoring. Imipramine is no longer recommended for enuresis because of the risk of fatal overdose.

Sleep Breathing Disorders in Children

Apneas of prematurity present in infants younger than 37 weeks as central, obstructive, or mixed apneas for at least 20 seconds, accompanied by hypoxia and low heart rate. Mixed apneas are the most common of all. Apneas are more common in low birth weight infants, especially those who weigh less than one kilogram. Apneas of prematurity results from immaturity of the respiratory center, but concurrent illnesses like anemia, hypoxia, infection, and medications raise the probability of developing this apnea. Apneas of prematurity usually resolve with maturity and rarely occur in full-term infants. Apnea persisting beyond 37 weeks is called *apnea of infancy*. Central apneas are more common than obstructive apneas after age 37 weeks and are associated with hypoxia, fluctuating brady-tachycardia, and arousals.

Obstructive sleep apnea in children may present with intermittent events of complete or partial decrements of the upper airway flow or as prolonged partial airway obstruction (hypoventilation) with oxygen desaturation and/or hypercapnia (carbon dioxide elevation). Obstructive sleep apnea affects 2-4% of children, especially preschool children (2 to 6 years) and adolescents. This disorder results from the combination of narrowing and hypotonia (muscle tone loss) of the upper airway pathways. Causes of obstructive sleep apnea in children include

- *adeno-tonsillar hypertrophy* (the most common risk factor for airway narrowing)
- craniofacial abnormalities, such as midfacial hypoplasia, micrognathia, mandibular hypoplasia, macroglossia
- hypotonia
- *Down syndrome* and *Prader-Willi syndrome* due to hypotonia, midline deformities, and obesity
- cerebral palsy with upper airway spasticity
- neuromuscular disorders with upper airway weakness
- obesity, especially in adolescents

At night, patients with obstructive sleep apnea may present with cyclic loud snoring, apnea, gasps, arousals, restless sleep, mouth breathing, sweating, and neck hyperextension. Children with hypoventilation may present with continuous snoring without apnea or arousals. During the day, children may present with cognitive and behavioral abnormalities, such as attention and concentration deficits, mood changes, hyperactivity, aggressiveness, poor school performance, and excessive daytime sleepiness, which is less common than in adults.

Figure 10-7 compares the main differences between sleep apnea in adults and children. If obstructive sleep apnea remains untreated, children may develop serious complications, such as failure to thrive, hypertension, cor pulmonale, and pulmonary hypertension. Polysomnography confirmation requires the presence of airway obstruction associated with gas exchange abnormalities (oxygen desaturation or hypercapnia), or sleep architecture changes to the different sleep stages with frequent arousals. *In children, one respiratory event (apnea or hypopnea) per hour is enough for the diagnosis of obstructive sleep apnea.*

The treatment of choice is adenotonsillectomy since adenotonsillar hypertrophy is the main cause of obstructive sleep apnea in childhood. However, the success of this surgery is not as high as originally thought, so surveillance for weeks after surgery for residual or recurrent obstructive sleep apnea is recommended. For children who have failed surgery or are not considered good candidates (craniofacial abnormalities, obesity, adolescence), continuous positive airway pressure devices are recommended. Positional therapy (sleep on the side) and oxygen therapy may help some children. Weight loss is recommended for obese children.

Autism: is a spectrum of disorders characterized by problems with communication, social interactions, and repetitive behaviors. Children and teenagers with autism suffer from several sleep issues. They struggle to fall or to stay sleep, sleep fragmentation, reduced total sleep time, refusal to compel with bedtime hours, and daytime behavioral consequences from sleep deprivation. Sleep hygiene, cognitive behavioral therapy, and melatonin can be helpful.

Insufficient Sleep in Children

Insufficient sleep (voluntary choice of sleeping less than required for age and individual, as mentioned above) is the most common cause of daytime sleepiness in children and adolescents. It often results from poor sleep hygiene, failure of parent to enforce sleeping hours, and/or side effects of medications. A third of high school students fall sleep in class at least once a week, and 14% of them come late or miss school due to oversleeping. In teenagers, insufficient sleep is also implicated in fatal car accidents. Daytime sleepiness resolves when patients choose to sleep longer. **Figure 10-8** shows common features of insufficient sleep in children and adolescents.

FIGURE 10-8 COMMON FEATURES OF INSUFFICIENT SLEEP IN CHILDREN AND ADOLESCENTS
• Increased need and effort to wake them up in the morning during weekdays • Sleeping two hours or more during weekends or vacations than during weekdays • Falling asleep at inappropriate times, as in school • Behavior and mood changes when they choose to sleep longer

Sleep Disorders in Women

During menstrual cycles, pregnancy, and menopause, women are affected by hormonal changes that can cause sleep disturbances. Contraceptives and hormone replacement therapy may also cause sleep problems. Insomnia is the most common sleep disturbance of women.

Menstrual Cycle and Sleep

The menstrual cycle is regulated by a series of hormones produced by the hypothalamic-pituitary-ovarian axis. The same hormones also work on brain areas that control the awake/sleep cycle, intertwining these two cycles.

A woman's menstrual cycle on average lasts 28 days and is divided into two phases, each one lasting about 14 days. The *follicular phase* starts the first day of menstruation (menstruation lasts 2-5 days) and lasts until ovulation. During the follicular phase,

FIGURE 10-7 DIFFERENCES BETWEEN SLEEP APNEA IN CHILDREN AND ADULTS

Main Features	Children	Adults
Excessive daytime sleepiness	Less common	Common
Hyperactivity	More common	Less common
Gender	Same	Men > Women
Cause	Adenotonsillar enlargement	Obesity
Polysomnography	1 or more apneas/hypopneas per hour	5 or more apneas/hypopneas per hour
Treatment	Adenotonsillectomy	Continuous positive airway pressure therapy

follicle-stimulating hormone and estrogen rise. The *luteal phase* lasts from ovulation until the first day of the menses. During this phase, the corpus luteum develops and releases progesterone and estradiol. Sleep disturbances are more common during the luteal phase, especially the week prior to the menses. At this time there is poor sleep quality, with a decline in total sleep time and sleep efficiency (percentage of the time spent in bed sleeping) that correlates with a fall in progesterone and estradiol levels. *Premenstrual syndrome* consists of emotional, behavioral, and physical changes a week before menses and may extend a few days into the menses. During this time, women complain of irritability, mood swings, fatigue, concentration problems, abdominal cramping, bloating, breast tenderness, headaches, and dysmenorrhea (painful uterine cramps). Premenstrual syndrome affects up to 18% of women. If symptoms become severe enough with depression and functional impairment, it is called *premenstrual dysphoric disorder*. Sleep disturbances are common with this disorder, including insomnia, fragmented sleep, non-restorative sleep, and excessive daytime sleepiness.

Pregnancy and Sleep

Pregnancy is often associated with sleep disturbances resulting from hormonal, physiological, and morphological changes. During pregnancy, estrogen and progesterone levels increase, leading to a narrowing of the nasopharynx by hyperemia, edema, and increased collapsibility. These changes in combination with an elevation of the diaphragm, reduction in respiratory capacity, and increased respiratory frequency cause a sensation of shortness of breath. Breast tenderness, nausea, headaches, nocturnal gastrointestinal reflux, constipation, leg cramps, lower back pain, abdominal pain, anxiety, baby movements, and nocturia (increased nocturnal urinary frequency) aggravate the sleep, especially during the second half of pregnancy.

During the last trimester, sleep becomes more fragmented, with more waking after sleep onset, increased amount of light sleep (NREM1), and reduction in total sleep time, NREM3 (slow-wave sleep), and sleep efficiency (**Fig. 10-9**). These changes, in combination with restless leg syndrome and sleep breathing disorders, which are more common at this time, lead to excessive daytime sleepiness and more naps.

If pregnancy is complicated by *preeclampsia* (hypertension, leg edema, and proteinuria), sleep becomes even more fragmented, with insomnia and daily sleepiness. The week prior to delivery and during the delivery day, especially at night, women develop sleep deprivation caused by a decline in progesterone levels and an increase in oxytocin levels.

The *postpartum period* is characterized by decreased placenta hormones (estrogen and progesterone) and mood changes. 50-80% of women develop "baby blues," sadness that lasts 2-3 weeks; 10-20% of women become depressed (especially with a prior history of depression) and 0.1% develop psychosis. Postpartum sleep becomes non-restorative, with sleep fragmentation, decreased sleep efficiency, decreased total sleep time, and decreased slow wave and REM sleep. Most of these sleep issues improve by the third month postpartum, but not to the pre-pregnancy status. Treatment of sleep disturbances in pregnancy focuses on sleep hygiene techniques rather than sedative drugs, which are generally not recommended.

Menopause and Sleep

Estrogen level declines during menopause, while follicle-stimulating hormone and testosterone rise. Pre-menopause, a period of years before menopause, is characterized by menstrual irregularities caused by hormonal fluctuations. Menopause is not diagnosed until there are no more menses for at least a year. During menopause, women have impairment in sleep quality and continuity. Sleep disturbances result from night sweats and hot flashes (vasomotor changes), which

FIGURE 10-9 COMMON SLEEP CHANGES DURING PREGNANCY AND POSTPARTUM	
First Trimester	Increased total sleep time due to naps Increased nocturnal insomnia Decreased NREM3 stage Increased waking after sleep onset Increased daytime sleepiness Increased daily naps
Second Trimester	Decreased total sleep time Increased awakening Decreased NREM3 stage Decreased REM sleep Increased waking after sleep onset
Third Trimester	Decreased total sleep time Increased nocturnal awakening Increased insomnia Increased NREM1 stage Decreased NREM2 stage Decreased NREM3 stage Decreased REM sleep Increase waking after sleep onset Increased daytime sleepiness
Postpartum	Decreased total sleep time Decreased NREM2 stage Decreased NREM3 stage Decreased REM sleep Increased waking after sleep onset

are more common at night. Pain, depression, anxiety, and sleep disorders (restless leg syndrome and sleep breathing disorders) often occur during menopause. The prevalence of sleep disorders in menopause is about 30-50%. During postmenopause, women continue with sleep complaints caused by comorbid medical problems, depression, and other sleep disorders such as insomnia, obstructive sleep apnea, and restless leg syndrome.

The most common sleep disorders seen in women are insomnia, obstructive sleep apnea, and restless leg syndrome.

Insomnia is more common in women than men. About one in three women have problems falling or staying sleep, or early awakening. Insomnia is more common during the late luteal phase of the menstrual cycle, third trimester of pregnancy, and menopause. Insomnia is more common in black women than Caucasians, in women of low socio-economic status, and in women with psychiatric illness, i.e. depression or anxiety. Women with insomnia have a fourfold likelihood of developing depression.

Treatment of insomnia focuses on sleep hygiene, reduction in fluid intake before bedtime, avoiding caffeine intake in the afternoon, avoiding substance abuse, and relaxation techniques. Naps are good, but not late in the afternoon, since this affects night sleep. Benzodiazepines and benzodiazepine-receptor agonists (i.e. *zolpidem, eszopiclone*), and melatonin-agonist drugs are often used. In the case of insomnia and depression, antidepressants with sleep-promoting properties are favored. In pregnancy, these drugs are not usually recommended since they may be associated with preterm, low birth weight, and neonatal respiratory depression. For more information about insomnia, refer to **Chapter 5**.

Obstructive sleep apnea affects 9% of women; however, women often remain underdiagnosed because of the atypical presentation. Women often complain of insomnia, morning headaches, fatigue, depression, and anxiety. Loud snoring, apneas, and excessive daytime sleepiness are less common in women than in men.

Women with obstructive sleep apnea are often obese. Obstructive sleep apnea is common during pregnancy, especially in the third trimester. It is also frequent during menopause and postmenopause when progesterone levels decrease (progesterone is a respiratory stimulant), especially if no hormonal replacement is used. However, hormonal replacement is not recommended in the treatment of obstructive sleep apnea in premenopausal women due to an increased thromboembolic risk, nor in menopausal or postmenopausal women due to lack of proven benefit. Women with obstructive sleep apnea have a significant risk of diabetes and vice versa.

The treatment of obstructive sleep apnea is the same for women and men (see **Chapter 3**).

Restless leg syndrome (**Chapter 9**) affects one in three women during the third trimester of pregnancy due to decreased iron. Restless leg syndrome is also common during menopause and postmenopause, though the etiology in these two groups of women remains unknown. Coexistence of various neurological disorders, obstructive sleep apnea, and aging may play a role in the genesis of restless leg syndrome during menopause and postmenopause.

The treatment of restless leg syndrome consists of avoiding triggers that worsen it, such as coffee, alcohol, and antihistamines. Iron replacement and drugs such as dopamine agonists, *pregabalin*, *gabapentin*, and opioids may be used. However, these drugs need to be used cautiously during pregnancy. For more information about restless leg syndrome see **Chapter 9**.

Sleep Disorders in the Elderly

Elderly people (older than 60 years) often present with sleep disturbances caused by non-restorative sleep and decreased total sleep time. One in two older people in America complains of chronic sleep problems from prolonged sleep latency, frequent awakening at night, decreased sleep efficiency, and earlier morning awakening. Reduced slow-wave sleep (NREM3), common in poor restorative sleep, accounts for much of the sleep disturbance in elders. Older people spend more time in bed, but less time sleeping. This lack of sufficient nighttime sleep leads to excessive daytime sleepiness, mood changes, cognitive impairment (poor memory and attention deficits), and falls. Elders who complain of daily sleepiness are also at risk to develop cancer, diabetes, hypertension, and heart disease. The sleep hours recommended for the elderly is 7-8 hours with a possible minimum of 5-6 hours and maximal of 9 hours. **Figure 10-10** shows common physiological sleep changes in the elderly.

Insomnia is the most common sleep disorder in the elderly. About 42% of older people present with

FIGURE 10-10 COMMON SLEEP CHANGES IN THE ELDERLY

Sleep Onset Latency	Prolonged
Sleep Efficiency	Decreased
Nocturnal Arousals	Increased
Excessive Daytime Sleepiness	Increased
Napping Time	Increased
Sleep Fragmentation	Increased
Total Sleep Time	Decreased
Deep Wave Sleep - NREM3	Decreased
REM Sleep	Decreased

FIGURE 10-11 COMMON SLEEP DISORDERS IN THE ELDERLY THAT CAUSE INSOMNIA

Primary Sleep Disorders	Secondary Sleep Disorders
• Sleep Apnea • Circadian Rhythm Disorders • Restless Leg Syndrome • REM-Behavior Disorders	• Painful, e.g. back pain • Neurological, e.g. Alzheimer's disease, Parkinson's disease • Psychiatric, e.g. depression, anxiety • Medical, e.g. heart or pulmonary disorders • Nocturia, e.g. prostatic hypertrophy, diabetes mellitus, overactive bladder, renal diseases • Stress • Medications • Toxins

sleep-onset and sleep-maintenance insomnia. Early awakening and frequent nocturnal awakening are also common. Women are more commonly affected than men. Insomnia may result from a primary or secondary sleep disorder. **Figure 10-11** shows common sleep disorders that cause insomnia in the elderly.

The treatment of insomnia focuses on cognitive behavioral therapy and sedative-hypnotic medications (antidepressants, antihistamines, anticonvulsants and antipsychotics), or their combination. Cognitive behavioral therapy is more effective than pharmacological therapy, and its effect lasts even after the therapy is over, as opposed to medications, which lose effect after discontinuance. For more information about insomnia, see **Chapter 5**.

Obstructive sleep apnea consists of repetitive cessation or decrease in nasal or oral airflow for at least 10 seconds during sleep, despite a continuing respiratory effort. The risk factors for obstructive sleep apnea in the elderly are the same as in younger patients: obesity, alcohol consumption, use of sedating drugs, smoking, family history of obstructive sleep apnea, and craniofacial abnormalities. Obstructive sleep apnea in the elderly is more common than in younger patients (men > women) and is more common in institutionalized patients than individuals living independently. 81% of institutionalized patients older than 65 years of age have an apnea-hypopnea index of 5 or more (diagnostic for obstructive sleep apnea), especially when patients are demented.

The clinical presentation of obstructive sleep apnea is similar to that in younger patients, with excessive daytime sleepiness. However, more unusual presentations of obstructive sleep apnea in the elderly are insomnia and nocturnal confusion. The chronic sleep deprivation from obstructive sleep apnea causes problems with cognitive function (impairment in memory and executive functions), manual dexterity, and balance.

Treatment of obstructive sleep apnea focuses on losing weight, eliminating trigger factors (drugs and alcohol), and use of continuous positive airway pressure. For more information about obstructive sleep apnea see **Chapter 3**.

Circadian rhythm disorder results when your internal sleep hour clock does not match with the external light-dark cycle. *Advanced sleep phase syndrome* is the most common circadian rhythm disorder of the elderly. These individuals go to bed earlier in the evening and awaken earlier in the morning than society's norm. There is normal sleep continuity with no daily sleepiness. If these sleeping hours do not affect the quality of life, no treatment is required. When sleep hours interfere with life activities, add lights in the evening hour to delay the internal circadian rhythm.

Restless leg syndrome and *periodic leg movements during sleep* are two different syndromes that often present in older individuals and are commonly confused as the same disorder. Restless leg syndrome is a clinical diagnosis based on four mandatory symptoms:

- There is leg discomfort, with an urge to move the leg(s).
- It improves with leg motion.
- It worsens with inactivity.
- It primarily presents at night.

Restless leg syndrome keeps the patient awake trying to relieve the unpleasant sensation, as opposed to periodic limb movement during sleep, which is involuntary, occurring while asleep. The diagnosis of periodic leg movement during sleep is made with polysomnography, which shows repetitive leg movement every 20-40 seconds during sleep, which might cause nocturnal awakenings. 80% of patients with restless leg syndrome have *periodic leg movement during sleep*, but the reverse is not true; only a minority of patients with periodic leg movement of sleep have restless leg syndrome. These two sleep disorders are often associated with obstructive sleep apnea. Restless leg syndrome commonly causes sleep onset insomnia, while periodic leg movement during sleep causes sleep-maintenance insomnia, both with excessive daytime sleepiness.

REM-behavior disorder is a parasomnia of older men, discussed in **Chapter 8**.

11

Sleep Medicine and Medical, Neurological, And Psychiatric Disorders

Medical, neurological, and psychiatric disorders are reciprocally related to sleep. Namely, sleep affects these disorders and these disorders affect sleep.

- *Medical disorders* associated with shortness of breath and pain cause frequent awakening and insomnia.
- *Neurological disorders* that affect brain regions involved in alertness and circadian rhythm result in insomnia, circadian rhythm disorders, and daily sleepiness.
- *Psychiatric disorders* are commonly accompanied by sleep disturbances, especially insomnia.

Drugs used to treat medical, neurological, and psychiatric disorders may worsen sleep. For instance, drugs used to treat insomnia can interfere with the sleep cycles, and the patient, while sleeping through the night, may not gain a refreshing sleep. The therapy for these sleep-related illnesses starts with optimization of the management of the primary disorder and subsequent treatment of the sleep complaints, if still present. This chapter focuses on the interrelationship between sleep and medicine, neurology, and psychiatry.

Medical Disorders and Sleep

Asthma

Asthma is a respiratory disorder that presents with intermittent cough, shortness of breath, and wheezing. It is caused by reversible bronchoconstriction and air hyper-reactivity to different stimuli. *Asthma worsens during sleep.* It presents with nocturnal cough, shortness of breath, insomnia, frequent nocturnal awakening, and early morning awakening. Daily sleepiness is common. Patients respond to steroids and bronchodilators. If asthma coexists with obstructive sleep apnea, *fixed-pressure continuous positive airway pressure (CPAP)* therapy is recommended.

Chronic Obstructive Pulmonary Disease (COPD)

Chronic obstructive pulmonary disease is a progressive irreversible respiratory disorder with destruction of the small airways and alveoli (emphysema) and airway narrowing with increased mucus secretion (chronic bronchitis). Patients present with chronic cough, shortness of breath, and, when severe, with hypoxemia (low oxygen levels) and hypercapnia (elevated carbon dioxide levels). Sleep is non-restorative due to delayed sleep onset, frequent awakenings, and insomnia. Some drugs used for chronic obstructive pulmonary disease worsen sleep, such as *methylxanthines*. Treatment focuses on stopping smoking, and the use of oxygen, bronchodilators, and steroid inhalers. 10% of patients with chronic obstructive pulmonary disease also have obstructive sleep apnea. This is called *overlap syndrome,* and its therapy consists of bilevel positive airway pressure (BiPAP).

Restrictive Pulmonary Diseases

Restrictive lung diseases are a group of disorders with diminished lung volumes and diffusion properties, caused by lung parenchyma disease and/or abnormalities of the chest wall structures (ribs, neuromuscular, pleura). Examples are interstitial lung disease, obesity, kyphoscoliosis, pregnancy, chest trauma, and certain neuromuscular disorders, including amyotrophic lateral sclerosis, myasthenia gravis, and other conditions with malfunctioning muscles. Patients present with shortness of breath on exertion and, when more severe, shortness of breath at rest. These disorders are commonly associated with obstructive and central sleep apneas with hypoxemia. Common complaints are poor restorative sleep, frequent awakening, insomnia, and daily sleepiness. Treatment focuses on the primary disorder, CPAP, and adaptive servo-ventilation (ASV) for obstructive and central sleep apneas respectively.

Gastrointestinal Disorders

Nocturnal gastroesophageal reflux is caused by the movement of gastric content back into the esophagus due to lower esophageal sphincter dysfunction. Patients present with nocturnal epigastric pain, prolonged sleep latency, frequent arousals, and insomnia. Diagnosis is confirmed with an *esophageal PH monitor,* showing a PH < 4 at the distal esophagus. Nocturnal gastroesophageal reflux is often associated with obstructive sleep apnea and may worsen asthma. Treatment consists of diet modification, head elevation, antacids, H2-blockers, and proton-pump inhibitors. Patients with nocturnal gastroesophageal reflux and obstructive sleep apnea improve the reflux symptoms with the use of CPAP (which elevates intra-esophageal pressure). Patients with duodenal ulcer disease complain of abdominal pain in the early part of sleep, correlating with an increase in gastric secretion. The pain, which may affect sleep, responds to H2 blockers, which help heal the ulcer. Sleeping on one's left side, rather than the right side, reduces GI reflux.

End-Stage Renal Disease

Patients with *end-stage renal disease,* with or without hemodialysis, often complain of sleep-onset insomnia, sleep-maintenance insomnia, early morning awakening, and daily sleepiness. The insomnia is caused by multiple factors, such as restless leg syndrome and periodic leg movements during sleep. Therapy of restless leg syndrome and periodic leg movements of sleep in chronic renal disease is no different than in non-renal patients with the exception that some drugs excreted by the kidney (e.g. gabapentin, pregabalin) require dose adjustments. Kidney transplantation is accompanied by resolution of restless leg syndrome.

Renal patients also commonly suffer from obstructive sleep apnea, especially the older population and those on chronic hemodialysis. Treatment with CPAP is the first therapy of choice. Changing hemodialysis to nocturnal hemodialysis, and renal transplant may also improve obstructive sleep apnea symptoms.

Cardiovascular Disorders

Cardiovascular disorders (*myocardial ischemia, congestive heart failure, cardiac arrhythmias, hypertension,* and *stroke*) are commonly associated with sleep disturbances, especially sleep breathing disorders. *Nocturnal myocardial ischemia* presents with chest pain, dyspnea, and palpitations that aggravate sleep. Nocturnal myocardial ischemia results from hypotension during sleep and the surges of the sympathetic nervous system during nocturnal awakenings. If nocturnal myocardial ischemia is complicated by sleep apnea, the ischemia worsens during the apnea. Polysomnography may reveal ST segment and T wave abnormalities in the ECG during the ischemia. Most myocardial infarctions occur between midnight and 6 am.

Congestive heart failure commonly presents with fatigue, dyspnea that affects sleep, and daily sleepiness. 40-80% of patients with congestive heart failure suffer from sleep breathing disorders, the most common being central sleep apnea, including Cheyne-Stokes breathing, and obstructive sleep apnea. *Cheyne-Stokes breathing* (discussed in more detail in **Chapter 4**) is a crescendo/decrescendo periodic breathing pattern that results from increased chemo-sensitivity of the arterial receptors to carbon dioxide levels. Sleep breathing disorders increase the mortality of patients with congestive heart failure, so if sleep apnea is suspected in a patient with congestive heart failure, polysomnography is required.

Refractory cardiac arrhythmias may result from the surges of the sympathetic nervous system during nocturnal awakenings in untreated obstructive sleep apnea with or without congestive heart failure.

There is a close association between sleep apnea and *hypertension.* 50% of patients with obstructive sleep apnea have hypertension, and 30% of hypertensive patients have obstructive sleep apnea, often undiagnosed. The risk factor for hypertension increases for each additional apnea per hour recorded. During sleep, systolic blood pressure normally decreases, a phenomenon called *"dipping."* Loss of dipping is often seen in obstructive sleep apnea, leading to increased cardiovascular mortality.

The primary therapy for patients with cardiovascular disease and sleep breathing disorders focuses first on improving the specific cardiovascular disorder. CPAP and BiPAP are recommended if sleep disturbances persist despite therapy of the primary disorder. They may reduce hypertension and improve cardiovascular symptoms and survival. Adaptive servo-ventilation (ASV) is indicated in Cheyne-Stokes breathing associated with CHF. *Stroke* is discussed below under *Neurological Disorders and Sleep*.

The Intensive Care Unit

Patients admitted to the intensive care unit commonly suffer from sleep deprivation (the *intensive care unit syndrome*). This results from the illness itself, as well as the intensive care unit environment (too much light, noise, drugs, and activity).

Diabetes

Diabetes contributes to sleep disturbances, especially painful diabetic peripheral neuropathy and renal disease. These cause sleep onset and maintenance insomnia, frequent awakening, and daily sleepiness. Diabetes is often associated with restless leg syndrome, especially when the patient has a neuropathy. Diabetes is more common in obese patients and therefore is often associated with obstructive sleep apnea.

Treatment focuses on adjustment of blood glucose, painkillers, CPAP for obstructive sleep apnea, dopamine agonists, and pregabalin or gabapentin for restless leg syndrome.

Thyroid Disorders

Hyperthyroidism presents with excessive sweating, tremors, heat intolerance, weight loss, palpitations, and cardiac arrhythmias. Common sleep problems are insomnia, nightmares, and night sweats. *Hypothyroidism* presents with weight gain, cold intolerance, and macroglossia (enlarged tongue). Obstructive sleep apnea and excessive daytime sleepiness are common.

Acromegaly

Increased secretion of growth hormone, usually from pituitary tumor, causes *acromegaly*, a disorder that includes orofacial anatomical changes. 60-80% of these patients develop obstructive sleep apnea.

Neurological Disorders and Sleep

Headaches

Migraine headaches are unilateral or bilateral headaches that last hours to days and are associated with light and noise sensitivity and nausea, with or without vomiting. Migraines affect sleep quality and duration. Conversely, sleep disorders worsen migraines.

Migraine patients often suffer from insomnia, sleep breathing disorders and circadian rhythm disorders. Abnormal sleep hours, oversleeping, and sleep deprivation are common triggers for migraine. Sleep may relieve or abort migraines.

Cluster headaches are intense unilateral headaches localized to the periorbital and frontotemporal region, lasting 30 min to three hours, with a same-sided, red, tearful, and droopy eye. Headaches occur during some months of the year and then remit and return. Patients with cluster headaches often wake up during the first REM period with an excruciating headache. There is a high incidence of obesity and sleep breathing disorders.

Hypnic headaches are a rare form of nocturnal headache in individuals older than 50 years. The headaches are unilateral or bilateral, brief (minutes), and awaken patients from REM and non-REM sleep. They are called "alarm clock" headaches because they are regular in onset (often between 1:00 and 3:00 AM) and confined to sleep.

Morning headaches are common in patients with obstructive sleep apnea, and obstructive sleep apnea should be suspected in obese men with headaches upon awakening that improve with the course of the day. The treatment targets the specific form of headaches. If a sleep breathing disorder complicates the headache, then CPAP may improve both the breathing disorder and the headaches. **Figure 11-1** shows common sleep findings encountered by patients with headaches.

Stroke

Stroke is defined as an acute neurological deficit of vascular origin in the central nervous system. Stroke frequently causes death and is a major source of neurological disability (cognitive and physical). Most strokes are ischemic; 10% are hemorrhagic. The risk factors for stroke are hypertension, diabetes, dyslipidemia, smoking, and atrial fibrillation.

Obstructive sleep apnea is a common and independent risk factor for stroke (ischemic and hemorrhagic), but is often neglected. 50% of stroke victims suffer from obstructive sleep apnea. The severity of obstructive

FIGURE 11-1 COMMON SLEEP DEFICITS IN NOCTURNAL HEADACHES

- Increased sleep onset latency
- Decreased total sleep time
- Sleep fragmentation
- Problems resuming sleeping if awakened at night

sleep apnea correlates with the severity and mortality of the stroke.

Obstructive sleep apnea is associated with intermittent nocturnal oxygen desaturation, fragmentation of sleep, reduction in cerebral blood flow, increased vascular inflammation, hypercoagulation, endothelial dysfunction, induction of hypertension, hypertension-induced left ventricular hypertrophy, and left atrial enlargement with atrial fibrillation. All these pathophysiological changes associated with obstructive sleep apnea increase the chance of having a stroke.

Stroke victims may suffer from central sleep apnea, including Cheyne-Stokes breathing disorder. Obstructive sleep apnea often persists after stroke recovery, but central sleep apneas disappear, suggesting that the central sleep apnea resulted from the stroke.

Treatment of obstructive sleep apnea with CPAP reduces the risk of stroke and mortality, and improves stroke neurological outcome.

Long nappers and sleepers, patients who take daytime naps longer than 90 min, and people who complain of poor sleep quality have an increased risk for stroke. Patients who survived a stroke and complain of sleep-awake disturbances (insomnia, poor sleep duration, restless leg syndrome, daytime sleepiness) are more likely to have another stroke.

Epilepsy

Epilepsy and sleep disorders are common comorbidities. Epileptics have frequent awakenings from sleep, with insomnia and daily sleepiness. Conversely, sleep disruption and deprivation are triggers for seizures. Most seizures, when they occur during sleep, occur during NREM2, and 20% of epileptics have seizures only during sleep. Antiepileptic drugs may affect sleep and cause insomnia and daytime sleepiness. Sleep breathing disorders are often seen in epileptics, and refractory epilepsy may result from untreated obstructive sleep apnea.

Dementia

Alzheimer dementia is characterized by a progressive brain degeneration that includes the alerting cholinergic pathway and the suprachiasmatic nuclei that control the awake/sleep circadian rhythm. Patients with dementia often present with sleep fragmentation and nocturnal awakening, insomnia, nocturia (frequent nocturnal urination), and excessive daytime sleepiness (**Fig. 11-2**). Sleep disturbances in Alzheimer's result primarily from dementia, associated depression, and comorbid sleep problems, including sleep breathing disorders, circadian rhythm disorders, and periodic leg movements of sleep. The severity of the sleep disorder correlates with progression of the dementia, risk of injuries, and decreased quality of life. Psychiatric symptoms often complicate the dementia and aggravate the sleep disruption. Benzodiazepines and antipsychotic drugs worsen excessive daytime sleepiness and increase the risk of injuries.

REM-behavior disorder is associated with *Lewy body dementia* and presents years ahead of dementia onset. *Sundowning* affects demented patients, presenting with wandering, restlessness, irritability, and agitation during late afternoon or early evening. Sundowning also aggravates sleep.

FIGURE 11-2 COMMON SLEEP DEFICITS IN DEMENTIA

- Decreased sleep efficiency
- Decreased total sleep time
- Increased sleep fragmentation
- Increased NREM1 stage
- Increased waking after sleep onset
- Decreased spindles on EEG
- Decreased K-complexes on EEG

Neuromuscular Disorders

Neuropathies, neuromuscular disorders (e.g. myasthenia gravis), motor neuron disorders (e.g. amyotrophic lateral sclerosis), and myopathies may weaken the respiratory muscles (intercostal and diaphragm) and oropharyngeal muscles, leading to restricted pulmonary function. This results in elevation of diurnal arterial carbon dioxide levels above 45 mmHg and nocturnal hypoventilation. *Nocturnal hypoventilation* is defined as elevation in nocturnal arterial carbon dioxide levels above 55 mmHg or a difference in arterial carbon dioxide levels of more than 10 mmHg between sleep and wakefulness. Patients with neuromuscular disorders are also at risk for sleep breathing disorders. These respiratory abnormalities cause sleep fragmentation and excessive daytime sleepiness, and, as the respiratory and sleep status worsen, so do the muscular fatigue and strength of the neuromuscular condition. Use of continuous positive airway pressure and bilevel positive airway pressure is recommended to treat sleep breathing disorders. To treat hypoventilation, non-invasive ventilation with bilevel positive-pressure ventilation is required (see **Chapter 4** for more information).

Multiple Sclerosis

Multiple sclerosis is an autoimmune demyelinating disease that randomly affects the central nervous system at multiple levels. If the damage affects the hypothalamic hypocretin cells, patients develop a *narcolepsy-type* picture. If the brain stem is mainly affected, patients develop *insomnia* (the most frequent complaint) and sometimes *REM-behavior disorder*. Sleep problems may result from multiple sclerosis symptoms, such as leg spasm, cramps, pain, immobility, and nocturia that affect the sleep continuity. Drugs used in multiple sclerosis may cause drowsiness.

Parkinson's Disease

Parkinson's disease is a neurodegenerative disorder caused by loss of dopamine-containing neurons of the brain stem substantia nigra. It presents with some combination of bradykinesia (slow movement), tremor, rigidity, and ataxia. Parkinson's disease is associated with multiple sleep disorders, such as restless leg syndrome, periodic leg movement disorder during sleep, REM-behavior disorder, obstructive sleep apnea, and nocturia. The REM-behavior disorder can precede the onset of Parkinson's disease motor symptoms for years to decades, and is considered one of the first manifestations of this illness. Parkinson's disease is also associated with insomnia, excessive daytime sleepiness, and nocturnal awakening, resulting from progressive stiffness and immobilization.

Fibromyalgia and Chronic Pain

Most *chronic pain syndromes* affect sleep quality and duration and result in insomnia, fatigue, and daily sleepiness. Drugs used to treat pain such as *opioids* can precipitate central sleep apnea. *Fibromyalgia* is a chronic diffuse muscle pain syndrome associated with specific tender points in the spinal region, arms, and legs. It is often associated with insomnia, non-restorative sleep, frequent nocturnal awakenings, and fatigue. Polysomnography displays a specific sleep abnormality called *alpha-delta sleep,* characterized by the interruption of slow waves (deep sleep EEG-pattern) by alpha waves (wake EEG-pattern). Sleep breathing disorders and restless leg syndrome are often present in these patients.

Brain Tumors

Brain tumors and *radiation therapy* that damages the hypothalamic hypocretin cells and brain stem reticular activating system may cause daily sleepiness. Conversely, tumors affecting melatonin cells of the pineal gland cause insomnia.

Psychiatric Disorders and Sleep

Depression

Depression is a mood disorder characterized by sadness, hopelessness, worthlessness, isolation, and anhedonia (loss of pleasure) with functional impairment (poor concentration, impaired decision-making, fatigue, and loss of appetite). Depression varies from mild to major episodes, sometimes with suicidal ideation. Depression can be a primary psychiatric disorder or may be secondary to substance/medication abuse or other causes. Most patients with depression suffer from insomnia (sleep-onset insomnia, sleep-maintenance insomnia, or early-morning awakening) (**Fig. 11-3**). However, some forms of depression, including *seasonal affective disorder* (winter depression), may present with excessive daytime sleepiness.

Treatment targets the depression, with discontinuation of toxins and medications that may cause or aggravate the depression, or involves psychotherapy and antidepressant drugs. Antidepressants with sedative properties given at night might be preferred, such as doxepin, amitriptyline, trazodone, mirtazapine, etc.

Anxiety

Anxiety causes significant functional impairment (social, interpersonal, work), with poor concentration, hyperarousal, and apprehension. Physical symptoms of muscle tension, muscle pain, fatigue, palpitations, diaphoresis (excessive sweating), and headaches are common. Sleep-onset insomnia, frequent nocturnal awakenings, non-restorative sleep, and daily sleepiness frequently accompany anxiety and decrease the quality of life. Treatment consists of selective serotonin reuptake inhibitors or serotonin-norepinephrine reuptake inhibitors. Benzodiazepines and antipsychotics are options only for refractory anxiety. Unfortunately, anti-anxiety drugs may trigger or worsen sleep disturbances.

Post-Traumatic Stress Disorder

Individuals exposed to life-threatening traumatic events (military combat, rape, childhood sexual abuse, serious illnesses, unexpected death of a loved one or a child, etc.) may re-experience the traumatic event months to years later; this is called *post-traumatic stress disorder (PTSD).* Patients have flashbacks, severe anxiety, anger, intrusion of images, and hypervigilance with startle reactions. These symptoms lead to avoidance of any situation

FIGURE 11-3 SLEEP CHANGES IN DEPRESSION

- Decreased sleep efficiency
- Decreased total sleep time
- Increased sleep fragmentation
- Increased NREM1 stage
- Increased waking after sleep onset
- Decreased NREM3 stage
- Shortened REM sleep latency

that might resemble the original traumatic experience, which in turn leads to interpersonal and work problems. Substance abuse (alcohol/drugs) is common. These patients often have difficulty falling or staying asleep, restless sleep, non-restorative sleep, and nightmares. Psychotherapy and selective serotonin reuptake inhibitor antidepressants may help with depression and mood changes. Discontinuation of alcohol and drugs is essential to improve sleep disturbances. *Prazosin*, an alpha-agonist blocker, helps to treat the nightmares. Treatment is similar to that in anxiety disorders.

Panic Disorder

Panic disorder is characterized by recurrent panic attacks. These are spells of intense fear that peak in minutes and last for about an hour. Patients present with autonomic activation, with dyspnea, diaphoresis, chest pain, palpitations, tremors, and numbness. They also have cognitive impairment and changes in behavior in a maladaptive way, where they live preoccupied with future attacks and avoid any circumstance that might trigger an attack. Substance abuse is common. Panic disorder may worsen other psychiatric disorders, i.e. anxiety, substance abuse, and psychosis. It is often accompanied by insomnia, frequent nocturnal awakening, and panic nocturnal arousals when the patient arises from the sleep. Treatment is similar to that in anxiety disorders.

Bipolar Disorders

Bipolar disorders present with major depression (see above), mania, and hypomania. *Manic* episodes consist of euphoria, hyperactivity, irritability, grandiosity, flight of ideas (rapidly moving from one idea to the next), and easy distractibility, for a period of at least a week. *Hypomania* resembles mania, but is less intense.

Bipolar disorders are divided into two subtypes: Bipolar I and Bipolar II. *Bipolar I* disorder presents with manic episodes and almost always with major depression and hypomanic episodes. *Bipolar II* presents with at least one hypomanic episode and one major depression episode, but no manic episodes. When depressed, the patient has insomnia and needs very little sleep during manic episodes.

Alcoholism

Alcohol consumption and withdrawal have significant effects on sleep continuity. Acute alcohol consumption at bedtime induces sleep, but when alcohol levels decrease 2-3 hours into sleep, patients develop frequent awakenings. Chronic alcohol consumption and its acute abstinence are often accompanied with frequent arousals from sleep. *Delirium tremens* (a state of agitation and hallucinations caused by acute alcohol withdrawal) is associated with severe sleep deprivation. Alcohol consumption aggravates other sleep disorders such as obstructive sleep apnea, parasomnias (sleepwalking and sleep terrors), enuresis, and nightmares.

Schizophrenia

Schizophrenia is a chronic psychiatric disorder characterized by delusions (fixed and unreasonable beliefs), hallucinations (perceptions without an external source), disorganized speech and behavior, decreased emotional expression present for at least a month, associated with significant functional impairment (work, interpersonal relationships, and self-care). Sleep-onset and sleep-maintenance insomnia or prolonged periods of sleep deprivation that end in excessive daytime sleepiness are common in schizophrenia. In addition, drugs used to treat schizophrenia may cause sedation and weight gain, which can lead to obstructive sleep apnea.

COVID-somnia

COVID-somnia refers to the sleep disorders associated with the COVID-19 pandemic. Long confinement at home is associated with dissolution of daily schedules, reduction in sunlight exposure, depression, anxiety, misuse and/or abuse of sleep medications, excessive use of electronic media and/or watching anxiety-provoking news, especially at bedtime. Patients develop insomnia, hypersomnia, circadian rhythm disorders, night terrors, disruption of sleep patterns, and excessive daytime sleepiness. Patients positive for COVID-19 develop sleep disorders from respiratory problems, fever/ malaise, medications, hospital admissions, worsening of pre-existing medical and/or sleep disorders, and home confinement.

Recommended treatment of COVID-10 somnia: Keep routine schedules as before COVID-19; sunlight exposure; avoid watching anxiety-provoking news and/or excessive alert-promoting blue light from electronics. Melatonin may help to treat circadian rhythm disorders. Zolpidem and benzodiazepines at night may treat insomnia, but avoid using them for more than a few days or weeks. If chronic insomnia or refractory insomnia develops, cognitive behavioral therapy is recommended. Treatment of sleep apnea with positive pressure is recommended for patients positive for COVID-19.

Suggested References

International Classification of Sleep Disorders, Third Edition, American Academy of Sleep Medicine

American Academy of Sleep Medicine. 2005. The International Classification of Sleep Disorders: Diagnostic & Coding Manual.

Berry R. 2012. Fundamentals of Sleep Medicine. Elsevier Saunders: Philadelphia (PA).

Kryger M, Roth T, Dement W. 2017. Principles and Practice of Sleep Medicine. Elsevier: Philadelphia (PA).

Lee-Chiong T. 2008. Sleep Medicine Essentials and Review. Oxford UP: New York (NY).

Gamaldo, C. Sleep Neurology. Continuum Life Learning in Neurology. American Academy of Neurology. August 2020.

Index